FACTS AND SPECULATIONS IN COSMOLOGY

The theory of the origin of the universe has advanced over time through observational evidence as well as through a lot of speculation. In this historical approach to cosmology, the authors review our present ideas on the origin and large-scale structure of the universe against the backdrop of our astronomical knowledge. They argue that the speculative element has become a dominant part of modern cosmology, showing how assumptions have been made and portrayed as confirmed facts.

This unique book gives not only a critical assessment of the big-bang theory, but presents a host of anomalous observations, and puts forward an alternative, controversial theory on the origin of the universe. A non-mathematical account, it contains analogies from everyday life so that readers can understand the concepts easily and follow the arguments presented. A thought-provoking insight into the evolution of cosmology for undergraduate students and general readers, this book shows that the mystery of the origin of the universe is far from being solved.

JAYANT V. NARLIKAR is Emeritus Professor at the Inter-University Centre for Astronomy and Astrophysics, Pune, India. Known for his work on cosmology, Narlikar served as President of the Cosmology Commission of the International Astronomical Union (1994–97). Other books by Professor Narlikar include *An Introduction to Cosmology* (now in its third edition, 2002) and *Current Issues in Cosmology* (with Jean-Claude Pecker, 2006), both with Cambridge University Press.

GEOFFREY BURBIDGE is a Professor in the Department of Physics at the University of California, San Diego. He is best known for his work on stellar nucleosynthesis and quasi-stellar objects. Professor Burbidge has been the recipient of several awards, most recently the Gold Medal of the Royal Astronomical Society with Margaret Burbidge. He co-authored *A Different Approach to Cosmology* (with Jayant Narlikar and Fred Hoyle, Cambridge University Press, released in paperback in 2005).

FACTS AND SPECULATIONS IN COSMOLOGY

JAYANT V. NARLIKAR

Inter-University Centre for Astronomy and Astrophysics, Pune, India

and

GEOFFREY BURBIDGE

University of California, San Diego

CAMBRIDGE
UNIVERSITY PRESS

CAMBRIDGE UNIVERSITY PRESS
Cambridge, New York, Melbourne, Madrid, Cape Town, Singapore, São Paulo, Delhi

Cambridge University Press
The Edinburgh Building, Cambridge CB2 8RU, UK

Published in the United States of America by Cambridge University Press, New York

www.cambridge.org
Information on this title: www.cambridge.org/9780521865043

First published 2008

Printed in the United Kingdom at the University Press, Cambridge

A catalogue record for this publication is available from the British Library

ISBN 978-0-521-86504-3 hardback

Contents

	Preface	*page* vii
1	Ancient cosmologies	1
2	The Greek epicycles	9
3	Reaching out to the Milky Way	28
4	Our position in the Galaxy	48
5	The world of galaxies	56
6	The expanding universe	75
7	Modelling the universe	88
8	What is the geometry of the universe like?	107
9	A universe without a beginning and without an end	119
10	The cosmological debate 1950–1965	136
11	The origin of the chemical elements	153
12	Cosmic microwave background	171
13	The very early universe	185
14	Dark matter and dark energy	207
15	An alternative cosmology	226
16	Unfaced challenges in cosmology	250
17	Epilogue	273
	Index	281

Preface

If you ask a well-read person what is his or her understanding about the origin of the universe, the reply is invariably: 'It originated in the big bang.' This is also the view of most professional scientists, that cosmologists are very close to their much sought after holy grail, namely the understanding of the origin and large-scale structure of the universe.

Major reasons for this view have come with the serendipitous discovery of the cosmic microwave background by Arno Penzias and Robert Wilson in 1964, and the discovery of the very small fluctuations in this by the teams led by George Smoot and John Mather in 1992. Both of these discoveries well deserved the Nobel prizes which have been awarded. The theory and observations, all based on the view that there must have been a big bang, have convinced all of the observers and most theorists that this model must be correct. However, this interpretation is in large part based on theory which has not been independently tested. In short, speculations are confused with facts.

Our purpose in writing this book is to share with the general reader our unease at the definitiveness that is often attached by professional cosmologists to their interpretations, when they are really based on speculations rather than facts. To underscore the circumstance that this is not the first such episode in the history of astronomy, we go through a historical sequence of the evolution of man's understanding of the universe. Time and again we find that there were majority views asserted with a great deal of firmness, views that had to be subsequently discarded when contradicted by facts. It is because of such examples, that distinguished physicists have expressed their derision of cosmologists in no uncertain terms. We cite two:

Lev Landau: 'Cosmologists are often wrong but never in doubt'

Max Born: '... modern cosmology has strayed from the sound empirical road into a wilderness where statements can be made without fear of observational check ...'

Many of our cosmologist colleagues would like to argue that all of the earlier examples in history belong to times when the observational support for cosmology was nowhere near as strong as it is today, so that they are on firmer ground in their assertions than their predecessors of bygone days. The fact is that those predecessors made exactly the same defence of their assertions.

The present theoretical structure of the standard big-bang cosmology rests on extrapolations of known physics to domains far beyond what is experimentally tested. Also the assertions about what happened in the very early universe do not rest on direct astronomical observations but on large extrapolations of what is actually observed. Given this speculative base of the present ideas, we feel that scope exists for alternative ideas that depart considerably from the big-bang paradigm. We have developed one such alternative model, the quasi-steady-state model. This predicts a cyclic universe with no beginning. We do not claim it to be perfect, but it is certainly good enough to be compared with the standard model. At present it is only rarely mentioned because of the hype associated with the big bang.

In this book we also draw attention to observations involving expansion and ejection in individual galaxies and clusters that are almost invariably ignored or disbelieved since they do not fit into the standard paradigm of the expanding universe. In the past the anomalous phenomena have played a useful role in shaping directions of physics and astronomy: they indicate that nature is holding out the hope of unravelling one more secret. Ignoring such data because 'it has to be wrong' is like throwing the proverbial baby out with the bath water.

We were coauthors with the late Fred Hoyle in a technical account of the present cosmology, entitled *A Different Approach to Cosmology*, published by Cambridge University Press in 2000. The book was quite well received by astronomers and physicists (but of course, not standard cosmologists!). This has encouraged us to write the present account so that the ideas may be shared with a wider group. We thank the Cambridge University Press for providing us with this opportunity.

We thank Peggy McCoy, and Vyankatesh Samak for help in preparing the manuscript. For line drawings we are glad to acknowledge help from Prem Kumar and for images and photographs, from Arvind Paranjpye.

1

Ancient cosmologies

The day of Brahma

The following ideas are found in ancient Indian mythological work, like the *Bhagavatam*:

The day or night of Brahma the Creator of the Universe is equivalent to 4 320 000 000 human years. This period is characteristic of the lifetime of one cycle of the universe, which is alternatively created and destroyed. Brahma plays the role of creator and Shiva the role of destroyer. In between Vishnu plays the role of preserver, who maintains the functions of the universe in each cycle.

This time scale is built out progressively in terms of cycles of *yugas*, which are very long periods. There are four yugas, with progressively diminishing states of the rule of the law or the truth: the *Satyayuga* having four out of four parts of truth, followed by the *Tretayuga*, having three parts truth and one part untruth, then *Dwaparayuga* with two parts of truth and two parts untruth, the last being *Kaliyuga* containing only one part truth against three parts untruth. Currently, according to these beliefs, the universe is passing through the Kaliyuga.

We will not go into details of how these time scales were arrived at but would like to emphasize that *they exist* in old literature, suggesting that the thinkers and philosophers of the day did think of time scales far in excess of the normal human lifespan. Moreover, they also had ideas on how these time scales were perceived to flow at different rates by different observers, as the following anecdote from the *Bhagavatam* shows.

A king called Kukudmi had a beautiful daughter Revati who had attracted several suitors for her hand. Anxious to choose the right Prince Charming, Kukudmi decided to take advice from no less an authority than Brahma himself. So he called on Brahma with his daughter. Brahma happened to be busy when Kukudmi arrived at his abode and so asked the visitors to wait a while. Shortly, after he finished his chores, Brahma called in his visitors and asked for the reason of their visit. When Kukudmi explained his mission, Brahma laughed and said: 'While you waited here a few minutes, several thousand years have elapsed on the Earth and the young men you had in mind for your daughter, exist no more. So let me suggest another name, of a prince who will be around when you get back to your kingdom.' So saying, Brahma suggested the name of Balarama.

This anecdote tells us that the *flow of time* was at a much slower rate at Brahma's location compared to that on the Earth.

The ancient ideas of space in these mythologies also make interesting reading. The Sanskrit word for the universe is *Brahmanda*, the *egg of Brahma*. This cosmic egg is supposed to include *all of the universe*, observed or otherwise. The stars and planets in the sky, the animate, vegetable and mineral world on the Earth and the unseen but imagined part *underground* were all supposed to be included in this cosmic egg.

Fig. 1.1. An artist's impression of the concept of the universe as conceived in Indian mythology. The entire creation is supposed to come out of a 'cosmic egg' or the 'Brahmanda'.

The Earth itself was believed to rest in a hierarchical structure on the heads of four elephants (in the four directions), which in turn rested on the back of a giant turtle, which rested on the head of a divine cobra.

As mentioned above, this large-scale structure was subject to periodic destruction and recreation. The cyclic idea also enters in the concept of *reincarnation* of living beings. The soul is indestructible, but the body is not. After death, the soul seeks another body to get into. In case you are worried how this reconciles with the increasing world population, the answer is that the bodies that the soul gets into need not be all human: the soul may enter into *any* being on the Earth . . . an insect or a bird or a fish or an animal. So, if one wants to check this hypothesis, one must keep track of the total number of living beings on the Earth!

Ra in ancient Egypt

The mythologies of ancient Egypt were different and equally colourful. Take the case of the Sun God Ra, who was supposed to travel through the sky in

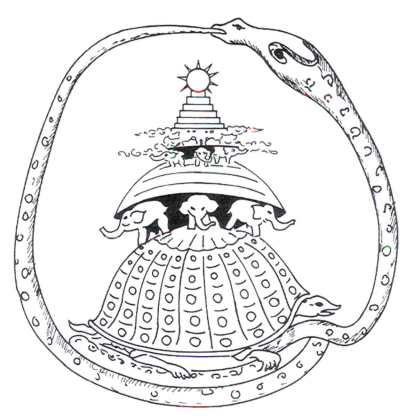

Fig. 1.2. Another concept from ancient Hindu mythology depicting the Earth as part of a cosmic hierarchy.

a boat called *Manjet*, or a *barque of a million years*. Many gods were supposed
to join him on this journey. The Manjet-boat sailed through twelve provinces,
representing twelve hours of daylight. At the end of the day Ra died and became
a corpse and sailed in another boat called the *Mesektet-boat*, meaning a *night-
barque* that sailed for twelve hours on water and appeared in the east in the
morning with a rejuvenated Ra.

One sees this idea described pictorially in ancient relics such as those in the
National Museum in Cairo. The Sun God is shown with a head like that of
a hawk with the solar disc on top, encircled by a cobra. Legends do not always

Fig. 1.3. Sun God Ra as given in ancient Egyptian findings.

depict a smooth sailing for Ra, in particular he has to frequently fight an enemy called *Apep*, in the form of a serpent. The god usually wins, except during solar eclipses when Apep swallows him.

Ra was believed to be the creator of light and other things found on Earth. He was supposed to have created the humankind from his tears and also the first couple, Shu and Tefnut, who were parents of the Earth and the sky. The solar disc itself was referred to as *Aten*, and Ra was supposed to reside in it.

The Norse world tree

In the Nordic civilization of northern Europe, the basic concept is of a world tree which carries the entire universe on its roots and branches. There are three levels in which nine 'worlds' reside.

At the upper level there are:

- Asgard (Æsir, the land of the gods)
- Alfheim (elves)
- Vanaheim (Vanir)

At the middle level we have:

- Midgard (men)
- Jotunheim (giants)
- Svartalfaheim (dark-elves)
- Nithavellir (dwarfs)

Finally the lower level houses:

- Muspelheim (fire, a bright, flaming, hot world in the southern region)
- Niflheim (the dead, the lowest level)

But the nine worlds and the world tree, Ygdrasil, were not there in the beginning. There are elaborate accounts of how they were created.

The Chinese mythology

The Chinese literature of the past is very rich and diverse. Here is a version of how it all began. First there was chaos everywhere and darkness. Then out of that emerged the cosmic egg (recall the Hindu notion of *Brahmanda*!). The egg harboured a sleeping giant of the name *Pangu* who was nurtured inside the egg for billions of years. When he grew and woke up, he stretched his body and came out breaking the egg. The lighter parts of the egg floated up to the sky

Fig. 1.4. The Norse 'world tree' was another hierarchical concept of the universe, as described in the text.

(Yang) while the heavier parts sank to become the Earth (Yin). The Yin–Yang duality comes in many aspects of Chinese thought.

Pangu liked this development, but he was afraid that some day the sky might fall on the Earth and so to prop it up, he grew and grew so that his feet were on the Earth and his head touched the sky. He grew at the rate of some 3 metres per day for 18 000 years and eventually increased the distance of the sky from the Earth to some 50 000 kilometres. Then he felt that they were reasonably stable and being very tired with his chore, he slept and never woke up.

When Pangu went into eternal sleep, his body was used to make the various parts of the universe. His eyes became the Sun and the Moon, his voice provided thunder and lightning, his breath led to the wind, his torso became the mountain while his arms and legs became the four directions. His flesh became soil, his blood the rivers while his veins became the roads. And so on and so forth, with the parasites on his body becoming the human beings of various races. Although he is no more, Pangu is still believed to control the weather according to his moods.

This story is ascribed to the Taoist writer Ko Hung who belonged to the fourth century AD, although its origin may be in South East Asia. It suggests the concept that the universe existed before man appeared in it.

From mythology to facts

Is there really a cosmic egg or a world tree housing the whole universe? Is there a Sun God as the Egyptians believed? What is Pangu up to now? These are mythologies that represent man's desire to 'know the final answer' about his origins, about the origins of what he observes and the extent of the universe around him. If he did not know for sure, he invented scenarios in which everything he saw fitted well, *given his ideas at the time*.

Do these ideas, graphic and detailed though they are, describe facts about the universe? The observations today will tell us that none of these ancient concepts make any sense. They are speculations not backed by facts. Yet they represent the most basic human aspiration to know the answer during one's lifetime.

This aspiration remains strong even today. . .despite the scientific requirement that all that we speculate about must have some basis in fact. Time and again we hear statements about science reaching the end of its goals, about the so-called end of physics, when everything fundamental is understood and details only need to be sorted out. Certainly, with the help of science we have begun to understand many of the details about the universe that baffled our ancestors. As Albert Einstein put it so graphically: 'The most incomprehensible thing about the universe is that it is comprehensible.' We know *what* makes the Sun shine. We know *how* the planets move around the Sun. We are able to figure out the interior of the atom, the smallest piece that makes up the various chemical elements. But how well do we know the universe as a whole?

In this book we follow up man's quest to find the answer to this question. Rather than present the modern picture in a cut and dried form, we shall try to present the developments and convictions that led to it. The progress towards understanding reality has not always been straightforward. But it is instructive to look with hindsight. We follow this historical path, although the main stress of our story will be on the more recent developments dating back a hundred years or more. Our bottom line is that although we know a lot about the universe, there is much more that we do not know.

2

The Greek epicycles

The central fire

School children have encountered the name of Pythagoras in their geometry books. His theorem about the square of the hypotenuse of a right-angled triangle being equal to the sum of the squares of its two perpendicular sides is a formidable one and plays a key role in Euclid's geometry, especially where measurements of distances are involved. Pythagoras was from Samos in Greece but left there for political reasons and settled in Kroton in Italy. He had to leave Kroton too because of opposition to his views and eventually died in Metapontum. He was a great thinker, a mystic and a religious reformer.

The followers of Pythagoras are known as Pythagoreans and amongst these was Philolaos who hailed from Thebes. The concept of the Sun–Earth system put forward by the Pythagoreans under Philolaos was, if anything, bizarre. Yet it gained acceptance in those times. What did it propose?

It stated that the Earth goes around a *central fire*, and not around the Sun! The Sun lay, according to this belief, outside the Earth's orbit. Why don't we see the central fire, some sceptics asked. The believers got around this difficulty by inventing a *counter-Earth* that also goes around the central fire but in an inner orbit in just such a way that it blocks the view of the central fire from the Earth! This argument silenced the critics, but only just. For they came up with another question: 'Why don't we see the counter-Earth?' To this the believers replied that Greece is on the other side facing away from the counter-Earth.

But eventually some sceptics sailed around to view from the other side and could not find the counter-Earth. So the hypothesis eventually died a natural death.

In this book we will have occasion to return to the hypothesis of the counter-Earth, for it very well illustrates what happens to a hypothesis that is not based on facts but is propped up by speculations. A wrong hypothesis

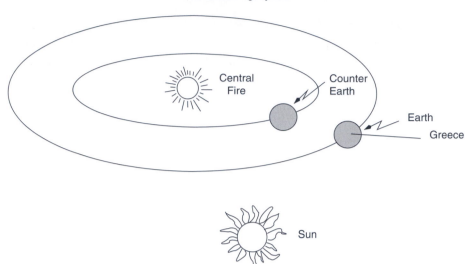

Fig. 2.1. The Pythagorean picture of the Earth and the Sun. The Earth moved around a central fire, whose view was blocked by a 'counter-Earth' that also moved around the fire. The counter-Earth could not be seen because Greece was on the other side.

inevitably encounters some factual evidence going against it. But, rather than abandon the hypothesis, a strong believer is tempted to invent an additional hypothesis to help cope with the discordant fact. This process does not end here. Often as more and more factual evidence accumulates, the supporters of the wrong hypothesis add on to it additional unproven assumptions to sustain it against the discordant facts. In most cases the entire structure becomes top-heavy and eventually collapses.

By contrast we will also come across hypotheses that are on the right track; in such cases additional evidence *supports* the hypothesis, which thus gains more credibility. Science advances through a trial and error process in which right as well as wrong hypotheses play their roles.

Let us examine another hypothesis about the cosmos that survived for nearly two millennia, until it could no longer continue under the burden of facts.

Epicycles and the geocentric theory

In contrast to the ideas described in the last chapter, the Pythagorean concept of the central fire and the counter-Earth marked progress towards the scientific approach in one respect. *The Pythagoreans proposed a hypothesis that could be tested.* The concepts of Brahmanda or the world tree did not have this property...they were entirely speculative. A scientific hypothesis must be testable and in principle, disprovable. That is, there should be a way of checking if the predictions made by the hypothesis are correct. If a prediction is made and checked to be correct, the hypothesis gains a point towards credibility. *However, it can never claim to be completely true!* It is always on probation and has to be ready for another, perhaps more sophisticated, test in the future. Who knows, there may be one test where it fails. In that case we either have to modify it into a better paradigm or replace it with another.

The Pythagorean ideas were eventually replaced by a more sophisticated framework based on the ideas of Aristotle (384 BC–322 BC). Aristotle was a pupil of the great Greek philosopher Plato and the teacher of another famous Greek, Alexander the Great. But Aristotle was a distinguished scholar in his own right, and his ideas on the nature of the universe were to dominate not only Greek civilization, but other civilizations in Europe and Asia as well. Indeed, in medieval Europe, Aristotle's ideas were so well established as to

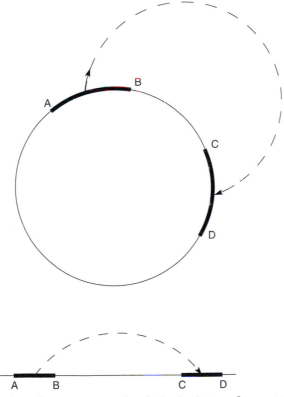

Fig. 2.2. To appreciate the symmetry of a circle, in the top figure, take any arc AB of the circle shown. If you imagine it being 'cut out' and placed on another part, like CD, of the circle, it will lie on it exactly. This feature is not shared by any other curve drawn on a plane, except by the straight line as shown in the bottom part of the figure.

become part of religious dogma. The key to Aristotle's ideas lies in his classi-
fication of different types of motion.

Aristotle distinguished two types of motion seen in the universe: *natural
motion*, which he supposed always to be in circles, and *violent motion*, which
was a departure from circular motion, and implied the existence of a disturbing
agency. Why circles? Because Aristotle was fascinated by a beautiful property
of circles which no other curve seemed to possess. Take any portion of a circle
(what we usually call a 'circular arc') and move it anywhere along the circumference: that portion will coincide exactly with the part of the circle underneath
it. (The straight line also has this property, but it can be considered as a circle
of infinite radius.)

In the jargon of modern theoretical physics, the above property is one of
symmetry. A one-dimensional creature moving along the circle will find all
locations on it exactly similar, there being no privileged position. In Chapter 7
we will find that present-day cosmologists employ similar symmetry arguments about the large-scale structure of the universe.

If you expose a photographic film for a long time to the light of stars in the
night sky, you will develop a picture of the sky full of circular arcs, these being
the trajectories of stars as they rise in the east and set in the west. The Sun and
the Moon also appear to follow circular paths in the sky. We thus see why
a prima facie case existed for Aristotle singling out circular trajectories as
nature's favourite trajectories.

There was a fly in the ointment, however! A handful of heavenly bodies did
not appear to follow circular tracks. Known as *planets* (meaning 'wanderers' in
Greek), these bodies appeared to defy Aristotle's edict of natural motion. Did
the planets possess some special power which enabled them to wander at will?

Those who answered this question in the affirmative were in one sense the
first astrologers, who argue that planets not only possess special powers but
exercise them on human destinies! It is ironical that even though the mystery of
planetary motion was fully resolved by Kepler and Newton, and these celestial
bodies were shown to be moving involuntarily under the force of gravitation,
belief in astrology is still widespread today.

Greek astronomers who did not go the way of astrology nevertheless missed
the chance of possibly discovering the law of gravitation. For, had they considered planetary motion to be of the violent kind, they might have been led to
wonder about the disturbing agency. Instead, they stuck to natural motion in
circles, and to reconcile the manifest lack of circular motion invented *epicycles*.

The primary hypothesis behind the epicyclic theory was that the Earth is at
rest in the universe, and that all heavenly bodies go around it. This theory
came to be known as the *geocentric* theory, since everything was viewed with

Fig. 2.3. The above photograph was obtained by keeping the camera aperture open all night. As stars move across the sky their images trace circular tracks. The Pole Star, however, stays fixed as a point because it lies very close to the axis of the spinning Earth. Credit: *Deepak Joshee, amateur astronomer, Pune, India.*

the Earth as the centre. The actual motion of bodies around the Earth may be circular, as it was found to be for stars, or it may be made up of two or more circles. Thus, in the simplest version, of one epicycle, a planet was considered to be moving in a circular path whose centre moved on another circular path around the fixed Earth. If this description was found inadequate to represent planetary motion and to forecast accurately the position of a planet in the sky,

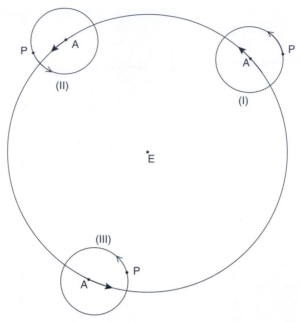

Fig. 2.4. This figure illustrates how the epicyclic theory works. Imagine a point A moving on the larger circle whose centre is at E. The smaller circle is centred at A and the planet P moves on it. In the figure we see three positions of A. When it moves from (I) to (II), the point P has moved also. The positions of A and P at a later stage are seen at (III). The Greeks felt that a planet moved like the point P around the Earth, imagined to be at E in this figure. Sometimes, in order to understand the apparent irregularities of the motion of a planet they had to introduce additional smaller circles mounted on one another. These circles came to be called *epicycles*.

more circles were added to the picture. Ptolemy carried the epicyclic theory to a high degree of sophistication in his classic book *Syntaxis*.

The Greek intellectual Aristarchus of Samos (*c.* 310 BC–230 BC) did not share the enthusiasm for these ideas. He argued that rather than let all the heavenly bodies go round in circles, it is simpler to imagine that the Earth itself spins about an axis. He further argued that the Sun is stationary and the Earth goes around it. Unfortunately his ideas had few takers and his writings were lost when the Alexandrian Library was destroyed by invading mobs.

Aristarchus had argued that as the Earth moves around the Sun, its location in space changes through the year. If you look at a street lamp from two different windows, it appears in the two cases with slightly different background. Aristarchus expected nearby stars to shift their position relative to the background of more distant stars if they were observed at six-month intervals. The effect, though real, was too small to be detected by the techniques of Aristarchus' time. So his theory did not receive support.

Fig. 2.5. This bust of Aristarchus (*c.* 310–230 BC) is proudly displayed in his native place Samos in Greece. The inscription reads: *Copernicus copied Aristarchus (1530)*. Photo by courtesy of Spiros Cotsakis of Samos. Credit: *Naresh Dadhich (private collection)*.

The next person to take issue with Greek astronomy was the Indian mathematician and astronomer Aryabhata. In his treatise entitled *Aryabhatiya* there is a verse which translates as follows: 'Just as a man moving in a boat downstream sees the trees on the river bank go in the opposite direction, so it is with stars which appear to move in the westerly direction when in fact they

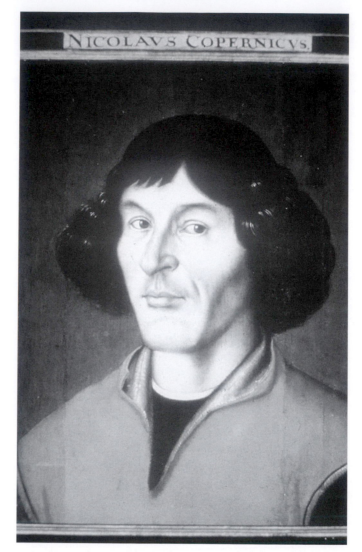

Fig. 2.6. Nicolaus Copernicus (1473–1543).
District Museum, Torun, Poland.

are fixed in space.' Aryabhata had thus discovered the fact of the Earth's revolution about its polar axis as the cause of the rising and setting of fixed stars. Thus the mystery of Aristotle's natural motion in circles was resolved as far as stellar motions were concerned.

But so deep-rooted were Aristotelian ideas in India in the fifth to eighth centuries that distinguished scholars who venerated Aryabhata in other respects either ignored this argument, or tried to reinterpret it in a way which did not conflict with the 'fixed Earth' hypothesis.

A more decisive challenge to the Greek dogma came in the fifteenth and sixteenth centuries, first from Nicolaus Copernicus and then from Galileo Galilei. Copernicus challenged the geocentric theory *in toto*. Not only was the Earth rotating about an axis, but it also travelled around the Sun, which was fixed in space; or so argued Copernicus in his volume *De Revolutionibus Orbium Celestium*. In this *heliocentric theory* (in which, as its name implies, everything is viewed with the Sun as the centre), epicycles are still needed, but the constructions are simpler than those given by Greek astronomers like Hipparchus and Ptolemy. Moreover, by arguing that the Earth and the planets go around the Sun, Copernicus paved the way for future theoreticians, who could look to the Sun for a possible cause of planetary motion.

The preface of the book by Copernicus is widely believed to have been tampered with. For, it reads very defensively and mild, considering the revolutionary nature of the main text. The reader is invited to treat the ideas as only a 'hypothesis' to be considered as an alternative to the standard belief. Perhaps the publisher was afraid of the hostile reception the book would receive and to protect himself, he changed the original preface that the author may have written. In any case, the book in its published form came to Copernicus's hands only as he was on his deathbed. So we do not know what his reaction would have been to the changes made.

Galileo and Kepler

Although Copernicus encountered considerable opposition to his revolutionary hypothesis, this was nothing compared to what Galileo had to suffer. Galileo not only supported the heliocentric theory; he also attacked the very basis of geocentric theory – namely, the ideas of Aristotle. And Galileo's objections to the Aristotelian school were not solely philosophical and conceptual; they were backed by experimental demonstrations. The famous experiment in which he dropped different kinds of bodies from the leaning tower of Pisa is but one among his many demonstrations. This experiment was reportedly performed to convince the Aristotelian supporters that all bodies, heavy or light fall with equal rapidity. The Aristotelians believed that the heavier bodies would fall faster. Normally in those days such issues would have been debated endlessly with verbal arguments. Galileo brought experimental demonstration into the picture for the first time.

While we admire Galileo for his acumen for experimentation, we should also give him credit for use of the telescope in astronomy for the first time. In 1609, the telescope had just been invented and while the military found immediate use for it to view enemy forces, so did Galileo for observing distant

Fig. 2.7. Galileo Galilei (1564–1642).
Louvre, Paris: Cabinet des Dessins.

cosmic objects. Here too his discoveries with the new gadget were not popular since they did not appear to fit in with the then current dogma. For example, he detected four satellites going around Jupiter. How could things go around another planet when everything was supposed to revolve around the centre-piece, Earth? The Moon was shown to have craters while the Sun was seen to have black spots (now known as sunspots) on it. How could creations of God have such 'blemishes' when the whole creation was believed to be a mark of

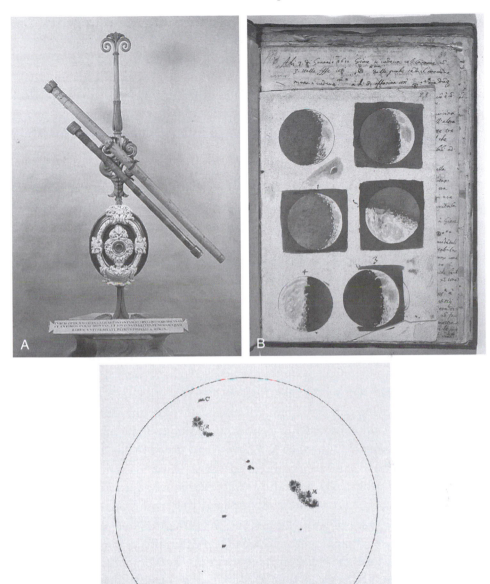

Fig. 2.8. A) Galileo's telescope. Museo di Storia della Scienza, Florence. B) Moon with craters seen from Galileo's first telescope as sketched by him. C) Picture of sunspots, first seen by Galileo and recorded by this sketch.
Images B) and C) courtesy History of Science Collections, University of Oklahoma Libraries; copyright the Board of Regents of the University of Oklahoma.

perfection? So the intellectuals shied away from the telescope which they regarded as nothing but magic or witchcraft.

Galileo's book *Dialogue on the Two Great World Systems* concerning two rival theories is a masterpiece of literary presentation of scientific arguments. It has three characters, one a supporter of Aristotle, another putting forth Galileo's views and the third a neutral person. It is shown that the second person always wins the argument, sometimes backed by experimental demonstration. The book indirectly heaped ridicule on the Aristotelian Establishment.

When the Establishment could take no more of this, it reacted in the only way it knew. Galileo was subjected to an inquisition, for arguing against the established tenets of religion. In the end, he recanted, but privately he retained his convictions about the correctness of the heliocentric theory and his objections to Aristotelian ideas to the end of his life.

Would Copernicus and Galileo have fared any better in modern times, in our so-called enlightened age of science? Today there are no inquisitions, and the Establishment is no longer constituted from the religious authorities, but, as far as radically new ideas are concerned, the situation today has not really changed. We will elaborate on this point at a suitable point towards the end of this book.

It is interesting to note in this context that Tycho Brahe, one of the greatest observational astronomers of his time, shared the geocentric view of the Establishment. He had an excellent observatory at Uraniborg in Denmark, where he conducted observations to establish that the Earth is at rest after all! Tycho later moved to Bohemia, where in 1601 he engaged a young assistant by the name of Johannes Kepler to work towards his goal of disproving the Copernican view through detailed observations. Tycho died in 1601, however, leaving behind all the wealth of his observations for Kepler to analyze. Kepler undertook the task in his characteristically meticulous manner, and after some twenty-five years of further observations and data analysis, came to the opposite conclusion, that Copernicus was right after all.

In fact, Kepler took the heliocentric theory considerably further than Copernicus and Galileo. He found, for example, that the planets moved around the Sun not in circles (or in circles on circles, and so on), but in ellipses. The elliptical orbits of all the planets had the Sun as a common focus. Kepler also discovered empirically the manner in which a planet moved along its orbit. These discoveries were summarized by Kepler in three laws of planetary motion.

Mathematically it is possible to describe the planetary motion in elliptical orbit as made up of a series of epicycles. In principle, an infinite number of epicycles is needed; but in practice, a small finite number gives a good approximation, because the orbits are nearly circular. Had the planetary orbits been considerably more elongated than they actually are, both

Fig. 2.9. Tycho Brahe (1546–1601) seen with the instruments in his observatory in
Uraniborg, Denmark.
Image courtesy History of Science Collections, University of Oklahoma Libraries;
copyright the Board of Regents of the University of Oklahoma.

Ptolemy and Copernicus would have found it hard to sustain the epicyclic
construction.

So with Kepler came the end of the idea of epicycles that had dominated
astronomy for nearly two millennia, from the times of Aristotle. We will
nevertheless have occasion to use the word 'epicycle' to describe the practice
in science where to sustain a hypothesis against a series of new observations, it

IOANNIS KEPPLERI
Mathematici Cæſarei
hanc Imaginem

ΙGENTORATENSI BIBLIOTHECÆ
Conſecr.

Fig. 2.10. Johannes Kepler (1571–1630).
Image courtesy History of Science Collections, University of Oklahoma Libraries;
copyright the Board of Regents of the University of Oklahoma.

becomes necessary to add some extra assumption. This assumption may in turn demand another extra assumption; and so the original hypothesis picks up a series of additional assumptions like the circles upon circles that the Greek astronomers needed in their epicyclic theory to describe planetary motion. Finally, in the case of planets, the real solution lay in an entirely different direction. The basic shape of an orbit is an ellipse rather than a circle and it is the Sun rather than the Earth that occupies the key position on one of its foci.

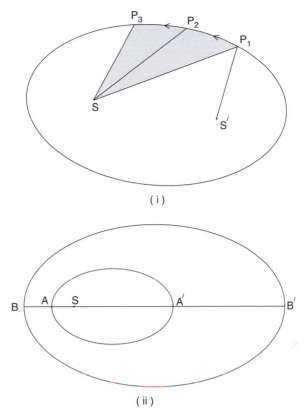

Fig. 2.11. Kepler's three laws of planetary motion can be explained using the above two figures. The first law shows the planet moving along the elliptical-shaped figure shown in (i) with the Sun at S, one of the two focal points of the ellipse. The other focal point is shown at S′. If you imagine a thread with its end tied to the two focal points and a pen moving around looping the thread at a moving point P_1, then P_1 will trace the ellipse. At what rate does P_1 move along the ellipse? The second law tells us that the line SP_1 traces equal areas in equal intervals of time. That is, if in the above figure, the time taken by the planet to move from P_1 to P_2 is the same as to move from P_2 to P_3, then the areas of the sections $SP_1 P_2$ and $SP_2 P_3$ are equal. In (ii) we see the explanation of the third law. Imagine two planets moving along ellipses of different sizes but both having the same Sun for the focal point S. The larger ellipse has the major axis BB′ while the smaller one has the major axis AA′. Take their ratio. Suppose that BB′ is r times as long as AA′. How much longer will the planet with the larger trajectory take to go round its ellipse, compared to the smaller planet? Suppose the ratio is s. Then Kepler's third law tells us that $s \times s = r \times r \times r$. For example, if BB′ is 4 times larger than AA′, then the planet moving on the larger ellipse will take 8 times as long to complete one round of the Sun. This is the law that tells us why, for example, the Martian year is longer than the Earth year.

So it may happen that the correct hypothesis provides an entirely different perspective on the situation, a perspective that is, at first, hard to accept.

The aftermath

Galileo died in 1642, and in the same year was born perhaps the greatest physicist the world has seen to date. Isaac Newton was destined to take the Keplerian achievement to its logical conclusion. Kepler had discovered the

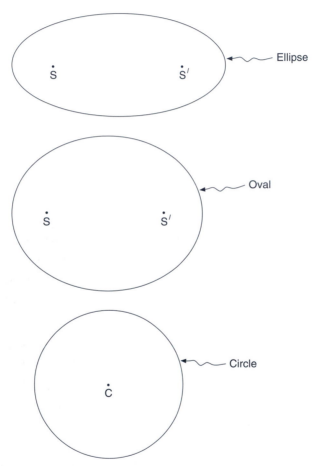

Fig. 2.12. Mathematicians tell us that there are many different types of 'egg-shaped' figures, like the ellipse and the oval shown at the top which look like an elongated version of the circle shown at the bottom. In the case of the ellipse, the sum of distances of any point on the curve from the two focal points S and S′ is the same. In the case of the oval (in particular, the classic Cassini's oval), the product of the two distances from two focal points is the same. When the focal points coincide, both the oval and the ellipse change into the circle. Kepler's achievement was to show that the planet moves around the Sun in an ellipse. The question was, why?

real pattern behind the motion of the planets. The question that he had been unable to answer was *why* the planets move in this way. Although the ellipse is a well-described geometrical figure it does not have the symmetry of the circle that had prompted Aristotle to assert that nature prefers circles. So, why an ellipse? Why not an oval or a spiral or some other curve?

Newton's great discovery was to show that ellipses result as natural answers if one assumes that the planets are attracted to the Sun by his law of gravitation. Kepler had already identified the Sun as the focal object in the motions of all planets. Empirically one could argue that *the Sun must have something to do with the motion of each planet.* But exactly what? Newton's law tells us that the planet is attracted towards the Sun with a force. Why should a force that attracts the planet *towards* the Sun make the planet go *around* it? Intuition suggests that the planet should fall *towards* the Sun.

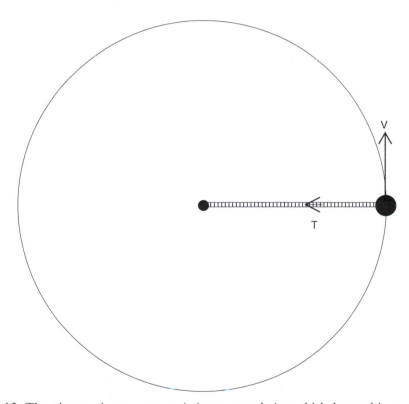

Fig. 2.13. The picture shows a stone tied to a rope being whirled round in a circular path. The rope is taut and there is a tension in it pulling the stone inwards. This tension causes the stone to change its direction of motion continuously as it moves in the circular path. The rate of change of motion is called *acceleration*. Thus the stone is accelerated towards the centre as it moves round.

This would be the case if the planet were initially *at rest* relative to the Sun. If instead it were initially moving around the Sun, that is, if it had a *transverse* velocity relative to the Sun, then it would continue to move around it. To see this tie a rope to a stone and whirl it round. The stone moves in a circle. Yet, if you ask what is the force acting on the stone, then you will find that the only force leading to this motion is in the rope and it is the tension of the rope acting towards the centre, that is, the hand holding the rope. Thus we see that *even though the force is radially inwards, the stone has transverse motion.*

This may appear counter-intuitive, but can be understood by another of Newton's basic laws, his second law of motion. This law relates the force acting on any piece of matter (like the stone in the above example) to the *acceleration* of the piece of matter. Now what is acceleration? It is the rate at which the velocity of the piece is changing. If you are travelling in a car along a straight road and your speedometer reads 60 kilometres per hour, your velocity is 60 kilometres per hour along the direction of your path. The velocity needs not only the *magnitude* but also the *direction* to specify it. Now suppose the speed limit sign says that you can go at 70 km per hour. Seeing no traffic you press the accelerator pedal to increase the petrol supply to the engine. Your car moves faster and soon your speedometer reads 70 km per hour. If you have a particularly powerful car, you will achieve this increase in speed sooner than someone with a moderately powerful car. So the acceleration measures not only the change of velocity but also how soon it is achieved. In the above example, if it took you one minute to increase the speed from 60 to 70 km per hour, your acceleration is 10 km per hour *per minute*, or since there are sixty minutes per hour, it is $60 \times 10 = 600$ km per hour per hour. It is directed along the path you are travelling.

What will happen if you drive a car along a circular track? Even if you keep your speed fixed, say, at 60 km per hour, you are still accelerating. . .because your direction of motion is constantly changing. So you need to apply the brake and accelerator suitably to keep on track. How much is the acceleration this time? To find out we need to know how rapidly your direction is changing. Suppose your track has radius 10 kilometres. For a circular track the rule is simple. Multiply the speed by itself and divide by the radius of the track. A little dynamical calculation tells us that in this case the acceleration is as much as 360 kilometres per hour per hour. *And, more importantly this is directed not along the track but perpendicular to it, directed towards the centre of the track.* It is because change of velocity is taking place like this that the car manages to stay on the circular track. If there were no such force acting on it it would have shot off in a straight line in the direction tangential to the circle.

Returning to Newton's law of motion, the force acting on a given piece of matter will cause it to accelerate (in the above car example, the pressing of the accelerator pedal to release more petrol generates force to propel the car forward with greater speed). However if the piece has greater mass, that is, if it has a greater amount of matter, it will require greater force to generate the same acceleration. The force that accelerates a tiny sports car will not be so effective in accelerating a massive truck.

Newton's law of gravitation lays down the rule that the force of attraction of the Sun on a planet is in proportion to the mass of the Sun as well as the mass of the planet. Further, the force is less and less effective the further the separation between the Sun and the planet. Thus, if the distance between the two were to increase to twice its original value, the force would drop fourfold in intensity; if the distance were increased tenfold the intensity would drop a hundredfold. The rule for drop in value is 2×2 in the first case and 10×10 in the second. For this reason, this rule is known as the 'inverse square law'. (The square of a number is obtained by multiplying the number by itself.)

Using this law and his law of motion described above, Newton was able to calculate how a planet would move under the gravitational influence of the Sun, if it had an initial velocity not in the direction towards or away from the Sun. And he found that the typical orbit would be an ellipse with the Sun at one of its two foci. Not only that, but the way the planet would move on this ellipse would be *exactly as per Kepler's laws.*

This resolution of the age-old question 'How do planets move?' is a triumph of the scientific approach of a basic theory describing a variety of motions of different planets exactly as per the observations. As mentioned earlier, the apparent lack of any rules in planetary motions had led many people to believe in astrology in ascribing special powers to planets. There remained no basis for such beliefs once Newtonian laws demonstrated how planets move. Nor was there any reason to build epicycles upon epicycles in describing planetary motions in the (excessive) belief in the universality of circular motions, as the followers of Aristotle had advocated.

The sequence of ideas from Copernicus to Newton through Galileo and Kepler represents a success story of man's attempts to understand the universe around him. But the quest did not end there…we shall look at the next landmark in the odyssey in the following chapter.

3

Reaching out to the Milky Way

Successes of the Newtonian framework

The Newtonian revolution resolved the puzzle of the planets. However, remembering what was described in the last chapter, one may ask the question: Wasn't the Newtonian theory just another exercise replacing one speculation, that of epicycles, by another, the inverse square law? Before proceeding further, it is advisable to dwell on this question at some length, only because it clearly separates what are genuine advances in science from pure speculations.

Newton designated the inverse square law as a 'universal' law; that is, it was not confined to just the Sun and the planets but was applicable to *all pieces of matter*. In the picture we see a statue of Newton sitting under an apple tree with an apple fallen in front of him. The legend of the apple is well known. Although it is apocryphal, with no authentication by any of Newton's official biographers, it makes a point that is worth pursuing further. There was an apple tree in Newton's garden and it is accepted that he often sat under it contemplating. It is also part of history that during the years 1665–66, when the plague drove people from cities to villages, Newton also left Cambridge and spent the period at his native home in Woolsthorpe. It was here during the two 'magical years' that Newton discovered several important results on which today's physics is founded.

The universality of Newton's law is indicated in the picture by the Moon and the apple, both of which are subject to Earth's gravitational attraction. The Moon goes around and around the Earth while the apple has fallen down vertically. The Moon has transverse velocity which is why it goes around. Recall that in the previous chapter we had a similar situation with regard to planets going around the Sun. We also saw that a stone tied to a rope and whirled around, moves in a circle, although the force acting on it is directed towards the centre of the circle. The apple has no transverse velocity and so it

Fig. 3.1. Isaac Newton shown sitting with the apple fallen at his feet. The falling apple is supposed to have inspired him with the idea of the force of gravitation, by which the Earth attracts the apple. The Moon shown in the picture suggests that the Moon is also attracted by the Earth by the same force. Credit: *Photo taken at IUCAA Arvind Paranjpye.*

falls straight towards the centre of the Earth. *But the motions of both these objects are governed by the same laws of motion and gravitation.*

Can we test this statement scientifically? With suitable apparatus we can measure the acceleration of the apple towards the Earth. This is approximately 9.8 metres per second per second. Then we can measure the radius of the Earth, say it is 6400 kilometres. Then applying Newton's law of gravitation we can determine the mass of the Earth, using the fact that it is producing the above acceleration at the above distance. For the Moon also, we can use astronomical observations to determine the time it takes to go around the Earth once. Say this is 29 days. Then we can measure its distance from us, say 380000 km. Using these facts we can work out the Moon's circular velocity around the Earth and then by the prescription given in the previous chapter, the acceleration of the Moon. Next, using the fact that so much acceleration is produced at the above distance from Earth, we can, using the inverse square law determine the mass of Earth. *The remarkable fact is that both these methods agree in giving the same answer.* Of course, the result is not so remarkable once we recognize that the Newtonian law applies equally well to the apple as it does to the Moon.

Now contrast this situation with the epicyclic theory. There, to understand each new observation about the planets we had to introduce a new epicycle. Here on the other hand, once the claim is made that the inverse square law applies to all material bodies, it becomes open to testing *any number of times*. Newton's law was so tested and came out on top every time. We will briefly mention two examples.

Halley's comet

Comets have been known for a long time as objects with a long tail that come into the sky from afar and return there after going around the Sun. Essentially a comet makes a U-turn about the Sun. For a long time comets were considered harbingers of bad news in the form of terrestrial catastrophes, deaths of important personages, defeats in wars, etc. Superstitions apart, the question remained: What made the comets come and go in this fashion? It was Edmund Halley an astronomer, a contemporary of Newton and one of his few friends who solved the puzzle.

Because of their 'portents' cometary visits have been recorded in various civilizations. Studying such records, Halley noticed in particular that cometary visits had taken place in the years 1456, 1531, 1607 and in his lifetime in the year 1682. Although chroniclers had talked of each visit in isolation relating to their lifetimes, Halley noticed a periodicity in these times, of the order of 75–76 years. Even in the year 1066 there had been a comet sighted and associated with King Harold's defeat at the hands of the invader William the Conqueror at the battle of Hastings. This also fitted in with this periodicity.

So Halley argued that these visits were not of different comets but *of the same comet* that came again and again because of the Sun's attraction. In

Fig. 3.2. Three pictures of the sighting of Halley's comet. The first one, immortalized on the Bayeux Tapestry depicts the sighting in the year 1066 AD by King Harold in England. The second one shows the comet seen over the river Thames in London in 1759 AD, the first sighting after Halley predicted the comet's visit seven decades earlier. Painting by Samuel Scott (source not traceable). The third picture is a photograph of the comet taken during its last visit in 1986. The comet takes approximately 76 years to travel around the Sun. *(Courtesy of the Indian Institute of Astrophysics, Bangalore, India.)*

short, just as a planet goes around the Sun in an elliptical orbit, so does a comet except that its ellipse is highly eccentric, that is, stretched out long with its two foci very far apart. (As the foci come closer together, the ellipse becomes more and more like a circle; it becomes a circle if the two foci coincide.) On the basis of the data available during the 1682 visit, Halley made calculations and predicted that the comet would next arrive in 1758. It did! Its arrival was expected on this occasion and greeted as a great success for mathematical predictions by Halley. The comet was in fact named after him. Halley's comet last paid us a visit in 1986 and its next visit is due in the year 2062. The firmness of the prediction and its accuracy, establishes our faith in the correctness of Newtonian physics.

The discovery of Neptune

Another example comes from the 1845 discovery of the planet Neptune. This example will also be cited later wherever the law of gravitation is used to infer the existence of matter that is not readily visible.

In the wake of Newton's laws and their success in understanding the motion of bodies in the Solar System, there was a feeling of 'well being' amongst the scientists towards the end of the eighteenth century. This was expressed by Frederick the Great who suggested that all important discoveries in astronomy had now been made and that nothing important remained to be discovered.[1] It was therefore a shock to the intelligentsia when William Herschel announced the discovery of a new planet which was subsequently named Uranus.

Herschel was a music master in Bath, a town in the west of England near the border with Wales. He played the organ in church services, conducted orchestras and gave organ recitals. But he was also an astronomer at night and observed the sky through the various telescopes that he had made. Besides this he studied books on mathematics and astronomy. But he was essentially an explorer at heart and instead of training his telescopes on known objects in the sky he would survey the sky to look for anything unusual. That is how in 1781 he came across Uranus which he first thought to be a comet, or a nebulous star. It took about a year to identify it as a planet lying beyond Saturn, until then (and from ancient times!) the furthest known planet. Lexell at St Petersburg demonstrated this fact which made Herschel famous.

However, as Uranus was being observed by astronomers in subsequent years, they began to discover a slight discrepancy in its orbit: it was not exactly following the trajectory that Kepler's laws would have predicted. Any possibility of observational errors was ruled out as time went along. What could be the reason? If Kepler's laws failed then that implied a failure of the Newtonian framework too, since that predicted the Keplerian results. Finally the discrepancy was resolved by John Couch Adams in Cambridge and, independently, by Urbaine Jean Leverrier in Paris. Both predicted that the cause of the discrepancy was a possible new planet in the vicinity of Uranus whose gravitational field was disturbing the Uranus orbit. Using Newton's laws they could calculate and find the possible location of the perturbing planet. If they

---

[1] It is curious that such a feeling appears towards the end of a century: we shall have occasion to discuss such pronouncements made towards the end of the nineteenth and twentieth centuries!

Fig. 3.3. John C. Adams, Urbaine J. Leverrier and Johann G. Galle, the principal
players in the discovery of the planet Neptune in the mid 1840s.
School of Mathematics and Statistics, University of St Andrews, Scotland.

(and Newton's laws) were correct then the planet should be seen in that direction.

Adams was the first to inform leading observers in England, the Astronomer Royal, George Airy and the Director of Cambridge Observatory, James Challis. However neither took a lead in actually looking for the new planet in the calculated direction. In the meantime, Leverrier also arrived at similar conclusions and communicated these to the French astronomers.

The leading French astronomers also did not pay serious attention to this suggestion, so Leverrier then contacted the Berlin Observatory. There Johann G. Galle, a young astronomer did take note, made attempts to look for the object in the suggested direction and *found it*. This was the discovery of Neptune.

This news was announced from Paris and Leverrier was given the credit for predicting the position. It was then that the British Establishment woke up to realize that had they taken note of Adams' suggestion he would have received the credit. Nevertheless, Adams was subsequently given due recognition for co-predicting the new planet. In any case both the British and the French had to pay the cost of not listening to their young astronomers.

As a postscript to this episode, one may add that when Leverrier's message reached the Berlin Observatory, the Director was away on leave because he was celebrating his birthday. So Galle, a young astronomer acted in his place. One wonders whether had the Director been present on duty, he too would have ignored the request to observe the new planet!

Herschel's map of the Milky Way

The discovery of Neptune was further striking proof that the mechanical movements in the cosmos follow the Newtonian laws of motion and gravitation. Indeed, the eighteenth century saw the consolidation of Newtonian mechanics and a rise in the belief that even the large and distant bodies of the universe follow laws of science. Perhaps nothing can demonstrate this attitude better than the following story about Laplace. The French mathematician and astronomer Pierre Simone de Laplace carried out extensive calculations to apply Newtonian mechanics to the entire Solar System as known at the time. This included planets as well as their satellites moving under one another's gravitational fields as well as that of the Sun. Laplace wrote a five-volume work on this topic under the title *Mecanique Celeste*, which he presented to Napoleon. It is said that after glancing through the highly mathematical descriptions, Napoleon commented: 'Monsieur Laplace, you do not mention God anywhere!' Laplace is said to have replied: 'Sire I had no need of that hypothesis.'

But as mathematical descriptions of astronomy became far more sophisticated than the early Greek attempts, observational techniques also improved so that the astronomer's attention was now shifted from the planets to the more distant objects, the stars. Herschel was primarily responsible for extending observational enterprise and human interest beyond traditional planetary astronomy. Despite his own contribution to planetary astronomy by the discovery of Uranus, Herschel perceived that the real

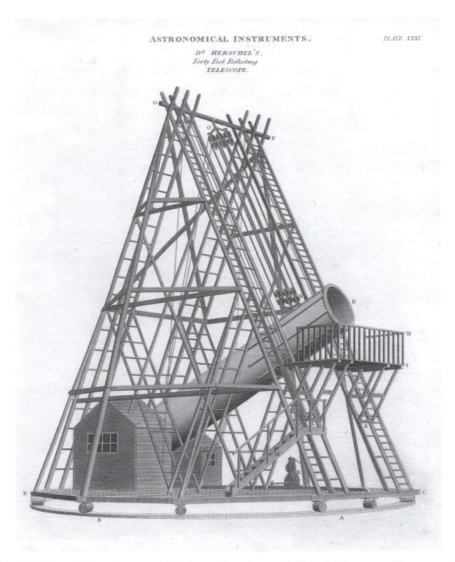

Fig. 3.4. Herschel's telescope with the main mirror of 48-inch diameter. However, the telescope was commonly known as the 40-foot telescope from the length of the tube. Source not traceable.

challenge existed in understanding stars and their distribution in space. And to meet this challenge he needed bigger and better telescopes.

To this end Herschel took on the role of a pioneering instrumentalist and he built telescopes with apertures 18.7 inches and 48 inches. (*Aperture* of a telescope is the diameter of its primary lens or mirror.) Naturally he needed financial support for this enterprise and was fortunate in having royal patronage. King George III was a great admirer of Herschel (the discovery of a new planet had certainly made Herschel's reputation) and extended full royal support to his observational efforts. The King paid a sum no less than £4000 for this reflector. When Herschel's 48-inch telescope, more commonly known as the 40-foot telescope because of the length of its tube, was ready, the King is said to have invited the Archbishop of Canterbury to look through it, with the remark: 'Come, My Lord Bishop, and I will show you the way to Heaven.'

Using his surveys of the stars in different parts of the sky Herschel prepared a map of the Milky Way. The Milky Way is the whitish band across the sky straddling the stars. To imagine what it really is, think first of a large disc filled with stars. Next imagine the Sun (and its Solar System) also in the disc and

Fig. 3.5. This photograph shows the band of white associated with the Milky Way. By noticing such a band in a different direction Herschel concluded that we are inside the Galaxy of stars that are seen projected as a white band. Credit: *Arvind Paranjpye (private collection)*.

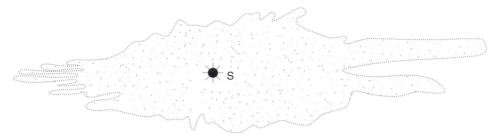

Fig. 3.6. The map of the Milky Way prepared by William Herschel more than two centuries ago. Notice that the Sun (at S) is placed very close to the centre of the system.

ourselves from Earth looking out. Just as someone in a forest sees a thick band of woods and leaves all around him, so would we see a dense band of shining stars all around provided we look in the directions lying within the disc.

Now if we were on the edge of the disc we would see the disc band only in one direction, not all around. Herschel, by examining the distribution of the stars, concluded that we are *within* the disc, not on the edge. Moreover, when he found the populations of stars in different directions of the disc to be the same, he concluded that the Sun is approximately at the centre of the Milky Way. The map of the Milky Way prepared by Herschel around the year 1785 remained more or less in force for a century and half. It was only during the twentieth century that astronomers realized that this perception was wrong. We will return to this issue in the following chapter.

We will, however, return to a more long-standing question. With improved observing instruments in the nineteenth century, astronomers could again address the question which had bugged them from time immemorial: Is the Earth fixed in space or does it move?

Does the Earth move?

Let us take a flash-back to the times of Galileo. During his inquisition, Galileo was asked to submit proof that the Earth actually moves around the Sun. Recall that Aristarchus of Samos (*see* Chapter 2) had argued similarly and produced as a test of his theory that the stars should apparently change direction in the sky when viewed at six-month intervals. No change, as estimated by Aristarchus, was found at the time and so his theory fell flat. Galileo offered a different proof. He argued that the phenomenon of tides in oceans showed that the Earth moved. He was drawing comparison with water being carried in a pot – as the pot moves, the water in it develops turbulence. So, Galileo argued, the oceans get stirred up because of the Earth's motion.

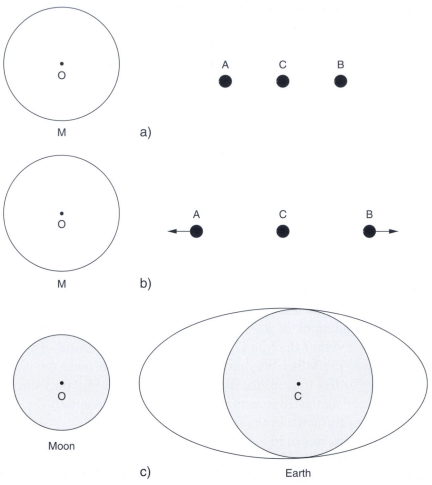

Fig. 3.7. The cause of tides. The tides on the high seas occur because of the law of gravitation. In a) we see three masses A, C, B at different distances from an attracting spherical mass M centred at O. Newton's inverse square law of gravitation tells us that all three masses will be pulled towards M, but in varying degrees. The force of gravitation is stronger at shorter distances and so mass A will be pulled more than mass C and mass C more than mass B. So, as shown in b) in relative terms, both masses A and B move away from C. Now apply this reasoning to the Earth and the Moon. As seen in c) the parts of the Earth towards and away from the Moon get pulled away from the central part, C of the Earth, just as in our example above of A and B moving away from C. This stretching force acts all over the Earth, but the rigidity of solid rock is able to withstand this and only the water surfaces, like the seas and oceans, are pliable and respond to it. Thus water 'jumps' up giving rise to tides.

In retrospect we know that the tides occur for a different reason (*see* Figure): the attraction of the Moon, and to a lesser extent that of the Sun, cause tides. But this reason, based on the law of gravitation was not known in Galileo's times. His inquisitors, who had made up their minds that he was wrong anyway, did not believe his explanation, right or wrong!

So to return to the question, how do we prove that the Earth is moving? There is one way of proving this by observing *aberration*. This method is based on the apparent change in the direction of the star observed from a moving platform, that is the Earth. To see how this method works consider the following example in daily life.

Suppose there is a pedestrian standing on the footpath under rain falling down vertically in a 'no-wind' condition. A car drives by at great speed. How does the driver in the car view the rain? He finds the rain lashing at the windscreen in an oblique direction. Why the obliquity? Because, just as a stationary pedestrian appears to the car driver to be moving in the opposite direction, so the rain acquires, relative to him, a horizontal component of motion in the opposite direction. This horizontal component added to the vertical component of velocity gives the rain an apparent oblique direction. And, as we can imagine, the obliquity will increase the faster the car moves.

Now replace the rain by rays of light coming from a star and the car by the moving Earth. If the Earth is supposed to move around the Sun, there will be an occasion when its velocity *towards* the star is largest and another when its velocity *away from* it is the largest. The *apparent* directions of the light rays on both these occasions would be different, as per the rain example. If the Earth were stationary, no such effect would be seen. This method was used by the astronomer James Bradley in the year 1728 to demonstrate that the Earth actually moves. The direction of the star *Gamma Draconis* shifted when observed at six-month intervals. From such observations we can estimate the Earth's speed relative to the Sun. It is approximately 30 kilometres per second.

This story will not be complete without a second reference to Aristarchus. He had overestimated the apparent change in the direction of a star when viewed from the Earth at two intervals separated by six months. The true effect is much smaller and could not be observed by Aristarchus' contemporaries. With improved observational techniques in the Herschellian era, this effect *could be* measured. The measured change in the direction to the star over a period of six months is called the *parallax* of the star. In 1838, Friedrich Bessel was the first observational astronomer to measure the parallax of the star *61 Cygni*. Its distance, as estimated from such measurements, is approximately 3.3 parsecs or 11 light years.

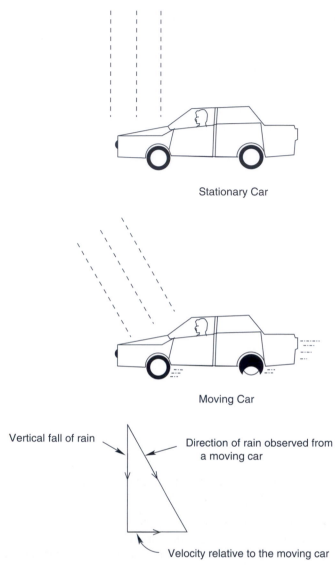

Fig. 3.8. Rain is falling vertically down as seen by a pedestrian standing on the footpath. To the driver of an automobile, however, it appears to fall in a direction inclined to the vertical. This is because, relative to him, the velocity of the rain has a horizontal component, which when added to the vertical component gives the inclined direction as shown in the adjoining figure.

Thus astronomers could not only assert that the Earth moves, but also acquire a method of measuring stellar distances. The measurement of distances of far-away objects in the universe which had baffled Aristarchus and his contemporaries, has continued to pose problems to the astronomer and

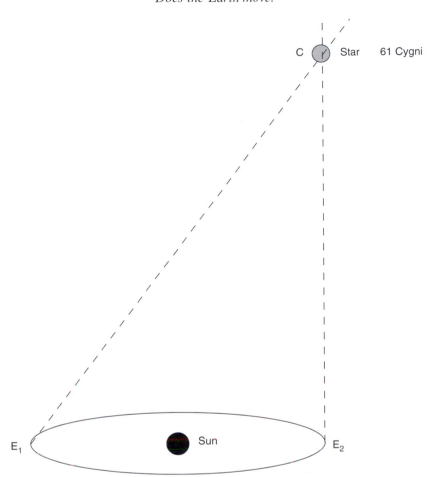

Fig. 3.9. The star at C is seen from two vantage points E_1 and E_2, the positions of the Earth six months apart as it goes around the Sun. The directions of the star as seen from these observing positions will be different as shown in the figure. This is known as the parallax effect. These measurements along with the knowledge of the size of Earth's orbit allow the astronomer to measure the distance of the star. The star 61 Cygni was the first one whose distance was measured this way. Notice that in the figure, the further away the star C is, the smaller the parallax angle E_1CE_2 subtended at the star by the two vantage points. In the times of Aristarchus, the instruments used were not capable of measuring small angles that stars subtend in this way and so his prediction could not be verified.

remains, even today, the one chink in his armour. We shall encounter in the rest of this book several instances of how uncertainties about distance have led to controversies.

As we saw above, the parallax method of measuring distance will work provided the distances are not very large. Starting from a few light year distances like that of *61 Cygni*, one may venture to longer distances of up to a few hundred light years. If one goes very far beyond this range the expected parallax will be so small that its measurement will involve uncontrollable errors. So for measuring such distances one needs some other, more reliable method.

The standard candle method of measuring distances

The photographic camera invented in the nineteenth century was a boon to the astronomer, as it enabled faint sources that were otherwise invisible to the eye, to be photographed. In a camera, the film can be exposed to the distant light source for a long enough time so that sufficient light is collected to form an image. In this way the camera serves as an ideal ally to the telescope in revealing the unseen universe. Not only stars but other, fainter cosmic objects have thus become targets for study.

Astronomers encountered examples of such faint objects in the sky. These are named generically as *nebulae*, and the word 'nebulous' in the English language, indicating a vague or hazy object or notion, has come from the name of such images. Notice that unlike stars, which appear as concentrated sources of light, these nebulae do not seem to have a sharp boundary; this suggests that they probably extend beyond what is seen in these pictures. Clearly a faster film and a longer exposure might reveal more. Indeed as the techniques of collecting faint images advanced, astronomers were able to view the universe more and more comprehensively. One can say that the universe extends far beyond what our eyes can see. The nebulae were to provide sources of controversy as astronomers sought to push back their observational horizons. We will encounter some of these in due course.

By collecting light from a star, the astronomer can measure what is known as its *apparent brightness*; apparent, because, the image does not contain the full information about the real *luminosity*, that is, the actual rate at which the star radiates energy.

An example with light bulbs will illustrate this issue. Suppose we view an illuminated light bulb of 10-watt power from a distance of 10 metres. We form a certain impression as to its brightness. As we move further away from the bulb, it appears to become fainter. From a distance of 100 metres it will appear very faint indeed. We say that the apparent brightness of the bulb decreases as its distance from us increases. At what rate does this decrease occur? To find out, the same experiment may be repeated with a 1000-watt bulb. Although intrinsically it is brighter than the 10-watt bulb, it too will appear fainter as we move away from it. However, we can ascertain from

Fig. 3.10. The Eagle Nebula shown as an example of nebulae, which are bright but diffuse clouds transcending stars. Credit: *NASA*.

several trials that its apparent brightness at a distance of 100 metres very closely matches the apparent brightness of the 10-watt bulb viewed from a distance of 10 metres.

This means that to compensate for a decrease in apparent brightness arising from a *tenfold* increase in distance, we need to boost up the luminosity of the bulb a *hundredfold*. The result may be generalized to what is commonly known as the *inverse square law of illumination*: the apparent brightness of a source of light falls off in proportion to the inverse of the square of its distance from the observer. Or, to put it differently, if we have two sources of light, source *A* being ten times further away than source *B*, then for them to appear equally bright to the observer source *A* must be *10^2 (= 10×10)* times as luminous as source *B*.

There is a simple way of understanding the inverse square law of illumination, which is illustrated by the adjoining figure. Here we have a light bulb *A* emitting radiation equally in all directions. Such a source is called an *isotropic* source. Let us assume that *A is emitting L units of energy per second; L is called* its *luminosity*. With *A* as the centre draw a sphere *S* with radius *r*. What is the area of the surface of *S* ? Some exercise of high school geometry will give the answer as $4\pi r^2$, where π is often approximated by the fraction 22/7. It is the same constant which when multiplied by the diameter of a circle gives its circumference. Taking this approximation for π we conclude that a sphere of radius 7 metres will have a surface area equal to 616 square metres. However, let us concentrate on the sphere of radius *r*. Imagine an observer *O* located somewhere on the surface of this sphere. *O* has a detector whose surface receives light from *A*. How much energy from *A* will come per second per unit area around this observer? This quantity will define the apparent

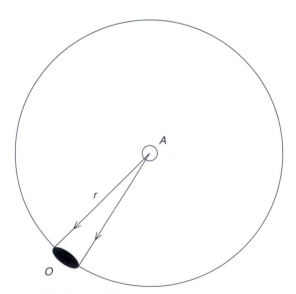

Fig. 3.11. The source of light *A* emits energy equally in all directions. It is therefore distributed evenly over the sphere centred on *A* and having radius *r*. Thus an observer *O* anywhere on this surface will find the same energy crossing unit area of the sphere around it. Since the total area of the sphere is $4\pi r^2$, a fraction of the total energy emitted equal to $1/4\pi r^2$ comes through the unit area around *O*.

brightness of the source. Since all points on the sphere are treated equally in the sharing of A's energy, and the area occupied by them all is $4\pi r^2$, the amount of radiation from A coming across unit area will be equal to the luminosity of A divided by $4\pi r^2$. The share received by O therefore comes *down* in proportion to r^2, that is, it drops down as the inverse square of r. The quantity $L/4\pi r^2$ is called the *apparent brightness* of A.

What applies to light bulbs applies also to stars. If we observe two stars A and B, and find that A looks much fainter than B, what can we conclude about their distances? *If* we know that A and B are stars of equal luminosity, then we can say that A is further away than B. But if we do not have this extra information then, naturally, we cannot make such an assertion. For example, A and B could be at the same distance, with A much less luminous than B. As it turns out, the stars that we can observe with the naked eye are not necessarily the nearest stars. By and large, they are more luminous and distant stars. Some of the really close stars are intrinsically so faint (i.e., with such low luminosity) that we cannot see them without telescopic aids.

Normally, with the help of telescopes and such modern light detectors as the charge coupled device, or the CCD, the astronomer is able to measure the apparent brightness of a source. If measurements of the distance of the source are also possible, then the astronomer can estimate the luminosity of the source. This is done by simply inverting the result obtained just now: multiply the observed apparent luminosity per unit area by $4\pi r^2$, where r is the measured distance to the source.

Let us apply this method to estimate the luminosity of the Sun. The Sun's distance from the Earth is about 150 million kilometres. The amount of light energy coming from the Sun across one square kilometre of area per second is about 1500 megawatts. That is, if we could convert all the solar energy falling on a one square kilometre area, we could use it to run a 1500-megawatt power station. So, by using the above method of calculation we estimate that the Sun has a luminosity of about 400 million million million megawatts! By terrestrial standards it is indeed enormous! But not by astronomical standards; for, if we do this exercise for many other stars, we find that the Sun lies about midway on the luminosity scale. There are stars a hundred times more luminous than the Sun and stars with only a hundredth of the luminosity of the Sun.

However, one rule of thumb that works reasonably well is that if we have stars of similar physical characteristics (e.g., mass, spectrum, etc.) then all of them will have roughly the same luminosity. So for such stars we have a standard luminosity, sometimes called the *standard candle*. Our earlier example of two stars A and B of the same luminosity can be used to compare their distances by comparing their apparent luminosities. If, in general, an astronomer can find a set of radiating objects all of the same luminosity then the standard candle method enables him to compare their relative distances. This method has been used extensively by astronomers for measuring distances of very far away objects. It fails, of course, if the standard candle assumption breaks down. We will describe some pitfalls of this kind in the next chapter.

Using the standard candle method astronomers were able to estimate stellar distances far beyond the limit accessible by the parallax method. This enabled them to form some idea of the vastness of the Milky Way.

How big is the Galaxy?

As astronomical measurement techniques improved through the nineteenth century, with photography revealing faint objects not previously seen, the concept of what constitutes our Galaxy took more concrete shape. The white band that we see across the sky which gave the Galaxy its name, was seen as the projection on the sky, of a disc-shaped stellar distribution, *as seen from inside*. There was believed to be a halo which contained other stars and stellar clusters *outside the disc*. There were dark patches in the white band which were believed to indicate *absence of stars*. There were nebulae, that is, whitish patches looking like clouds, both within and outside the disc that were believed to be gaseous regions lit by stars within. The disc itself showed the existence of spiral arms that gradually opened out from a dense galactic centre, arms indicating denser concentrations of stars compared to the gaps between them. The Sun along with its planets was widely believed to occupy the central region of the Milky Way. This overall picture was accepted till the dawn of the twentieth century.

This picture was built on the basis of such distance estimates as were then available to the astronomer; the methods of obtaining them, we have just outlined. How far was this picture correct? With hindsight today we can point out where it went wrong; but at the time the overall feeling was that the picture must be correct.

It is perhaps a human urge that during one's lifetime one should settle the age-old question: What is our universe like: how big is it; how old is it; how did it begin? Which is why succeeding generations of astronomers have been tempted to assert that the cosmological problem is all but solved.

At every age, however, pursuit of these goals, has depended on the state of the technology. Every human technology has its limits beyond which it cannot work. Galileo introduced a giant leap in the observing ability of mankind. But his 3–4-cm-aperture telescopes had their limits. Herschel drew on all the then available technology of optics and mechanical engineering to erect his 48-inch

(1.2-metre) telescope. But he had problems in operating this at its full expected capacity. History tells us that many of Herschel's important measurements were made with his smaller telescopes.

Telescopes, however large and sophisticated, have their limits in observing capability. Close to these limits their reliability becomes more and more questionable. The same applies to physical theories, tested under a specific range of circumstances, which are then extrapolated well beyond those limits by enthusiastic (and adventurous!) theoreticians. Gravity theories of Newton and Einstein have been applied well past the range of their actual physical verification. To give an example, the parameter 'α' measuring the effect of gravity on space in Einstein's general theory of relativity, is tested for values up to around one part in a million within our Solar System. For asserting the existence of a black hole one needs to visualize circumstances where this parameter approaches a value close to unity, that is almost a million times the tested limit.

Of course scientists are aware that extrapolations beyond the tested range, whether in observation or in theory, carry their baggage of speculation. Indeed it is through daring speculation that science advances. Newton's assertion that his law of gravitation was 'universal' was a brilliant but daring extrapolation. But what appears to be a speculation today may well be tested by a technology of the future. Through such tests the next advance could be made.

However, what is not often realized is that not all speculations are confirmed as facts, and until such vindication today's extrapolations are no more than speculations. It is when astronomers are interpreting data lying close to the limit of their instrumental ability that they are treading grey areas where controversies may arise. Likewise testing physical theories beyond their existing range can be fraught with problems of interpretation.

In the following chapter we will encounter controversies prevailing in the first quarter of the twentieth century. For reasons just discussed such controversies were inevitable. Indeed one can say that controversies are indicative of the health of the subject. It is when controversies are suppressed, as happened with Copernicus and Galileo that one should begin to doubt if all is well with the subject.

4

Our position in the Galaxy

How far away are stars?

We saw in the previous chapter that one way an astronomer estimates the distance of a star is by measuring its faintness. The standard assumption behind this method is that the fainter the star the further away it is. To some extent, this assumption is justified by the inverse square law of illumination, that we encountered in the last chapter. However, that law has its weaknesses, as astronomers realized to their cost. A couple of examples will illustrate the pitfalls.

Suppose we have a star *A* which is a hundred times more luminous than star *B*. Suppose further that both *A* and *B* are located at the same distance. Then to us *B* will appear faint compared to *A*, having a hundredth of the brightness of *A*. So, using our inverse square law of illumination, we will deduce that *B* must be ten times further away than *A*. The fallacy behind this conclusion is that the law we have used assumes that all stars are equally luminous. Had *B* been as luminous as *A*, then, of course, this conclusion would have been correct.

Often when looking at new species of objects, astronomers tend to spot the brightest ones first as they stand out amongst the rest. They then make the (mistaken) assumption that *all* objects of the same species are equally powerful. So when they estimate the distance of a typical object, they assume that it is looking faint because of its long distance. So they end up *overestimating* the distance of such an object.

The second source of error is demonstrated by the example of a nebula shown in the figure.

It shows a dark picture shaped like the head of a horse which gives the nebula its name. This represents a large region in our Galaxy, several light years across, containing many stars. The darkness is predominantly because of

48

Fig. 4.1. The Horse Head Nebula. The dark shape has nothing to do with any horse, but represents absorption of starlight by interstellar dust. Credit: *NASA*.

the absorption of light by dust. The photograph of an urban scene shown overleaf shows how dust particles in the atmosphere can produce attenuation of light.

So the horse's head does not mean that there are no stars in that region of the Galaxy. Rather it implies the existence of light-obscuring agents like dust particles. Careful photography using different wavelengths can assure us of this reality. Indeed, interstellar dust did not come to be recognized as a major obscuring agent until the early parts of the last century. Examples like the Horse Head Nebula are curtains between light sources like the stars and us the observers.

Imagine now what havoc dust can play with stellar distance measurements. Let us go back to our example of the two stars *A* and *B*. This time we assume that both are equally luminous and at the same distance away from us. However, suppose that there is a curtain of dust between us and *B*, a curtain that removes three quarters of the light from *B*. This means we will receive only a quarter of the light that left *B* and so we will make the mistake of assuming

Fig. 4.2. Absorption of light by fog and dust can obscure views on the Earth and make transport difficult. Here we see part of the Golden Gate bridge fogged out. Credit: *Keith Noto, Department of Computer Sciences and Department of Biostatistics and Medical Informatics, University of Wisconsin, USA.*

that *B* is apparently fainter than *A* by a factor of four. Using the inverse square law of illumination now would tell us that *B* must be *twice* as far away as *A* in order that it appears four times as faint as *A* ($2 \times 2 = 4$). In other words, we would interpret *B* to be further away than it actually is.

In general a light beam travelling through interstellar dust would be partially scattered from its original direction by dust particles and partly it would be absorbed by them. The result is less intensity in the original direction as the beam progresses further. This phenomenon is called *interstellar extinction*, and it combines the effects of both scattering and absorption.

Not realizing the significance of interstellar dust, astronomers of the nineteenth century overestimated stellar distances and so had erroneous ideas about the shape of our Galaxy. In some cases, the early telescopes were simply incapable of seeing far enough to appreciate the correct extent of the Galaxy. Which is why the picture of the Milky Way put together by William Herschel

Fig. 4.3. Milky Way band observed in different directions shows the light from stars in those directions. Because of their enormous distances it is not possible to spot individual stars in this band as points of light. (*Courtesy of NASA*).

(*see* Chapter 3) was very incomplete in terms of stars not seen at all, and based on wrong distance estimates of stars actually seen.

The galactic centre

The modern picture of the Milky Way, that is, our Galaxy, is seen in the above photograph which is put together by joining pictures taken in different directions. Imagine that we are seeing the Galaxy from inside and trying to put together a picture of how stars are distributed within it. The bright patches are stars not seen individually, but *collectively*. Rather, like a crowd of people photographed from a distance. In the latter picture we cannot make out individual faces or human bodies since the ability of our camera to see clearly is limited. Likewise photographs taken through the best of telescopes have a limit on the *resolution* of individual stars. The bright band of light that we see in a clear night sky as the Milky Way is in fact made of several hundreds of thousands of stars.

But no less important in the picture is the dark band that intersperses across the white band of the Milky Way. Now we know that it is not dark because of *absence* of stars but because of the *presence* of interstellar dust. What is the dust made of? The answer to this question is slowly being discovered. One clue to the solution is provided by the fact that how much of the incident light is scattered and how much is absorbed depends on three major factors: the material of the dust particle, its shape and size and on the wavelength of light falling on it.

But from what we have said so far, it is clear that unless one knows the full extent of the object one is looking at, it is hard to identify its centre. In his time Herschel had placed the Sun and its planetary system close to the centre of the Galaxy. This perception had not changed as late as the first decade of the twentieth century. The Dutch astronomer J. C. Kapteyn at the Mount Wilson Observatory used the techniques of astronomical photography and by 1920, had arrived at a revised model of our Galaxy. In this model stars are largely distributed in a plane, called the galactic plane. A careful study of the distribution of the stars showed that the Galaxy was shaped like a flattened spheroidal system with the diameter of the galactic plane being 40 000 light years and the extension in the perpendicular direction to the plane of around 8000 light years. This was believed to be the extent of our visible universe and came to be known as the *Kapteyn Universe*. It had the Sun slightly away from the centre, at a distance of some 2000 light years. This picture was not significantly different, qualitatively, from what Herschel had produced; quantitatively it had better measurements of stellar distances. However, as was made clear later, the role of interstellar extinction of starlight had not been taken into account and so the distances estimated were grossly incorrect.

Parallel to this development, during 1915–1919, a challenge emerged to the Kapteyn Universe through the work of Harlow Shapley, who had carefully measured the distances of a class of stellar systems called *globular clusters*. As its name implies, a globular cluster is a distribution of a large number of stars, some hundreds of thousands of them, that looks like a sphere. Shapley found that the globular clusters were uniformly distributed perpendicular to the galactic plane. However, within the plane they were concentrated along the direction of the Sagittarius constellation. From this observation Shapley felt

Fig. 4.4. J. C. Kapteyn and H. Shapley who were adversaries in a debate on where we are located inside the Milky Way. Credit: *Pencil sketches, Arvind Paranjpye.*

that the Galactic Centre (GC) lay in that direction, since one would expect a concentration of stars as we move towards the GC. He estimated the Sun's distance from the GC to be as much as 50 000 light years. This, again was a gross overestimate because the interstellar extinction was not allowed for.

Clearly therefore, the Kapteyn Universe was different from Shapley's picture of the Galaxy. Which was right? As happens in such cases, there was considerable argument and debate, with the majority siding with Kapteyn

Fig. 4.5. A globular cluster is a spherical-looking distribution of a large number (thousands) of stars moving in one another's gravitational attraction. More stars in the distribution congregate in the central region because of the force of gravity. Credit: *NASA/STScI.*

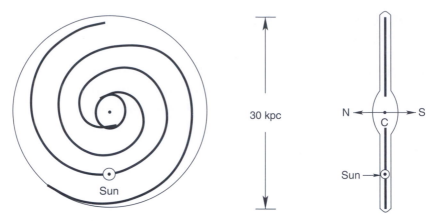

Fig. 4.6. The picture on the left shows a schematic view of the Milky Way galaxy seen face-on. It has spiral arms with the Sun seen on one of these. The picture on the right is the edge-on view showing the bulge in the central region. Notice that the Sun is not at the centre of the Galaxy.

and the minority going with Shapley. The situation was gradually resolved through the decade 1920–1930 and observations by Jan Oort, another Dutch astronomer and others gradually shifted the opinion towards the Shapley picture, although the distances estimated by Shapley had to be reduced to take account of interstellar extinction.

Jumping across several decades, we have today's perspective on our Galaxy: it is a disc-shaped distribution of stars with the Sun at a distance of around 30 000 light years from the GC. There is a small bulge near the centre with stars concentrated around it. The disc itself has a diameter of around 100 000 light years and is about 3000 light years thick and has a large halo surrounding it. The stars are distributed in the disc in a non-uniform manner, being concentrated along several arms spiralling out of the GC. For this reason, our Milky Way is classed as a *spiral galaxy*. The Sun, along with other stars goes around the GC. It would complete one revolution of its orbit in about 250 million years.

Looking at the history of how man has perceived himself as a denizen of the universe, we cannot help but note the irony of his demotion from a privileged position. The Greeks had man occupying the Earth as the centrepiece of the universe. This position gave way with the Copernican revolution which demoted the Earth as just one of the many planets going around the Sun. The Sun then came to occupy a position of importance and human ego may have been satisfied when Herschel accorded it a position near the centre of the Galaxy. Alas, that was not to be! With better information, astronomers showed that far from being at or close to the GC, the Sun is about two-thirds of the way towards the periphery of the galactic disc. Nevertheless, some satisfaction could still be salvaged by noting, what was the common belief a hundred years ago that our Galaxy, the Milky Way was the dominant, perhaps the only massive and large object in the universe.

As we shall see next, even this satisfaction was denied to the *Homo sapiens*.

5

The world of galaxies

Kant and the island universe

In the last chapter we showed that the astronomical perception of where we are in the universe got progressively better as the observing techniques improved. Each landmark of progress led to the destruction of a belief cherished by the majority not only of the human population but also of its intellectual component. Consciously or unconsciously, the human ego had assumed that we here on Earth are occupying the most important position in the universe. Even if our Earth is not fixed but goes around the Sun, and even if the Sun is not at the Galactic Centre but goes around it, can we not derive some satisfaction from the fact that we are part of the gigantic system of the stars, namely our Galaxy, nowadays called the Milky Way which is unique in the universe?

Although a hundred years ago it was tempting to think so, there were dissenting voices here too. For example, Immanuel Kant (1724–1804) had argued back in the eighteenth century for a universe in which the Milky Way was just one of the many similar galaxies floating in a vast universe like islands in a vast ocean. Known as the *island universe hypothesis*, this idea had few takers at the time. The mathematician J. H. Lambert (1728–1777) was one of them. He believed not only that the Sun is going around the Galactic Centre, but also in the existence of other galaxies lying well outside the Milky Way.

The references to these island universes sprang from the observations of the so-called nebulae. Unlike the stars that are shining points of light, the nebulae are diffuse and faint. One cannot make out from naked-eye observations even through large telescopes, where a nebula ends: it merges into the dark sky. The Messier catalogue of nebulae was initiated in the eighteenth century (1771–84) by Charles Messier in France. This was a first compilation of about one hundred nebulae of various kinds that had been seen by astronomers through naked-eye observing with telescopes.

In the nineteenth century, the technique of astronomical photography was developed and this provided immense help in the studies of nebulae. By exposing photographic plates for long periods it was possible to get clear images of nebulae, with much better detail than could be perceived by the human eye. Some nebulae from the Messier catalogue are now shown.

The first of these pictures is of the *Crab Nebula*, showing the debris left behind by a stellar explosion. The exploded star is called a *supernova*. The star was identified as the one that the Chinese astronomers and American Indians saw in 1054 AD. There is no record of it being seen in Europe, probably because there

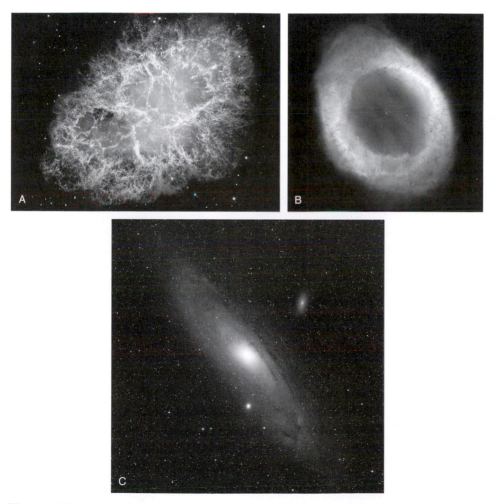

Fig. 5.1. Three nebulae are shown to illustrate the cause of the great debate in the second decade of the last century. The Crab and the Ring nebulae are in fact part of our Milky Way galaxy, whereas the Andromeda Nebula is outside this and much further away than the first two. However, the majority of astronomers until about 1920 believed that all such nebulae were part of our Galaxy. Credit: *NASA*.

was much more cloud. Also, the level of civilization was much lower in Europe than it was in China a thousand years ago. According to the Chinese, on July 4 of that year the star brightened so much that its light was sufficient for reading at night. For a few days the star was visible even during the day and it gradually faded away. What we see in the Crab Nebula is the remnant of that explosion.

The second picture represents a much more peaceful event! Called the *Ring Nebula*, it shows gaseous rings blown off by a star that has become too large in size and can no longer hold on to its outer parts. (We need to remind ourselves that the star holds on to all its body by the force of its own gravitational attraction. In a typical star, as it grows to a giant size, the force of attraction of the central region on the periphery is rather low and this can result in the boundary layers blowing themselves off.) The star itself can be seen at the centre of the doughnut-shaped ring and it is its radiation that lights the ring just as a planet is lit by a star. Not surprisingly astronomers call nebulae of this kind, planetary nebulae.

Both of these nebulae have come from stars and are part of the Milky Way. Their nature took some time to establish and only after the development of the physics of stellar structure and evolution, could astronomers understand what these nebulae are about. But what about the third picture? The object is known as the Andromeda Nebula because of the constellation in which it was found; the question is: *What does it represent?*

William Herschel had worried about nebulae of this type and wondered if they were gigantic star-systems lying far away. Later he came to link them with gaseous nebulae like the Crab or the Ring Nebula. Lambert on the other hand opted for the first alternative. Significantly, many of these nebulae seemed to lie away from the plane of the Milky Way. Why? Did that fact suggest that they were linked to the Milky Way in some way?

A supporter of Lambert's ideas, R. A. Proctor felt the opposite to be the case. He argued that if there was interstellar dust, it would absorb light from those nebulae whose line of sight was in the plane of the Milky Way. Hence we see only those that lie away from the galactic disc. An episode from real life will help to explain this argument.

One December evening, one of us was flying from Pune to Delhi. Although the Indian winter is mild compared to that in, say, northern Europe or Canada, the climatic conditions near Delhi can create thick fog making it difficult, sometimes impossible, for aircraft to land. On this occasion as the aircraft circled the Delhi airport prior to its descent, the passengers could see the road and motor traffic below quite clearly. Nevertheless, the pilot announced: 'The visibility is very marginal, and I have been advised against landing. We will be diverted to another airport...I will brief you on this decision soon.' As there

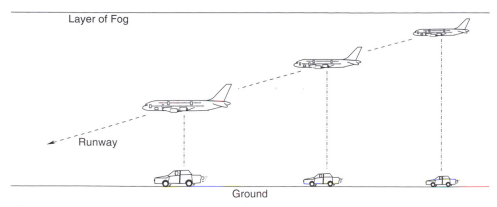

Fig. 5.2. The aircraft landing in fog requires visibility in the forward, nearly horizontal direction, where the fog extends over a long distance. So the pilot may not be able to land. In the vertical direction, however, the layer of fog may be thin enough to provide a passenger in the aircraft a reasonable view of the ground below.

was a howl of disappointment from the passengers, the question was asked as to why was the pilot finding landing problematical, when they could clearly see the roads below. The answer lies in the fact that the layer of fog is like a thick horizontal curtain. Its thickness in the vertical direction is not large so that light can penetrate through it and one can therefore see the traffic below. However, when landing, the pilot has to see in the forward direction which is almost horizontal and here the layer of fog is quite thick, too thick for him to be able to see through it.

Likewise, when looking through the Milky Way, we can see more clearly in the direction perpendicular to the galactic disc which has the absorbing dust but cannot see far *within* the disc. So astronomers would see only those nebulae which were *away* from the Milky Way plane. Factually correct though Proctor's view was, for a long time this remained a minority view not believed by most astronomers.

An indication of how the majority viewed the situation even as late as a century ago is shown by the following extract from a popular book by Agnes Clerke:

The question whether nebulae are external galaxies hardly any longer needs discussion. No competent thinker, with the whole of the available evidence before him, can now, it is safe to say, maintain any single nebula to be a star system of co-ordinate rank with the Milky Way. A practical certainty has been attained that the entire contents, stellar and nebula, of the sphere belong to one mighty aggregation, and stand in ordered mutual relations within the limits of one all embracing scheme.

Agnes Clerke,
The System of the Stars, 1905

That is, she is ruling out the possibility that any nebula like Andromeda could be a galaxy in its own right lying far away from the Milky Way. It was such a belief that led to the formulation of the Kapteyn Universe model in which the whole universe is seen to be centred around our Galaxy. However, like the other majority beliefs this also met its Waterloo in the fields of observation.

The observations of van Maanen

Astronomy, perhaps more than other sciences has been continuously sustained and has flourished on controversy. Controversies can serve as fresh stimulus to new observations, often going to greater levels of sophistication to resolve the question under debate. The island universe hypothesis of Immanuel Kant became the subject of one in the opening decades of the twentieth century. The majority view, so categorically stated by Agnes Clerke, began to receive more challenges as new information began to be available through larger telescopes and better instruments to collect data.

For example, like the Andromeda Nebula, there were several nebulae found with a spiral structure. Known as *spiral nebulae*, these were believed by some to be galaxies like ours with billions of stars. Heber D. Curtis, for example, had measured the distances of some of these to be far greater than the extent of the Milky Way. The distance of the Andromeda Nebula (M 31) came out to be half a million light years. Recall from the previous chapter that the extent of the Milky Way disc is around a hundred thousand light years.

One consequence was the debate between Harlow Shapley and Heber Curtis in 1920 in the Smithsonian Institution of Washington, DC. The debate was about the overall shape and size of the universe. Shapley had already 'demolished' the perception that our Solar System is close to the centre of the Milky Way galaxy. The view that he supported in the debate was that there was no likelihood that the spiral nebulae seen in astronomical observations were galaxies in their own right. Curtis supported the opposite view. At the time the debate hinged on the credibility one would attach to the observations of the Dutch astronomer working in the Mount Wilson Observatory, Adrian van Maanen.

Adrian van Maanen was a Dutch-born American astronomer from Sneek, the Netherlands. Of aristocratic parentage, he was educated at the University of Utrecht, where he earned his B.A. in 1906, his M.A. in 1909 and his Ph.D. two years later. His first position was at the University of Groningen. He chose astronomy as his career at a time when the subject was undergoing a rapid change for the better, with new technology for telescopes and instruments to be used with them. He took advantage of these facilities to the full. He moved to the USA to join the Yerkes Observatory in 1911 soon after its opening. He joined as a volunteer in an unpaid capacity to gain experience in astronomy. He moved to a paid position at the Mount Wilson Observatory where he was mostly involved with measurements of distances and motions of the stars. At the time the 60-inch reflecting telescope on Mount Wilson was the most advanced one in astronomy.

Most of van Maanen's working life was devoted to studying the motion of planetary nebulae and faint stars. He continued to draw upon technological improvements in observing techniques and his meticulous measurements formed the basis for future investigations by astronomers. He passed away in 1946. However, in an ironical twist of fortune, his observations of spiral nebulae served to prolong the debate in the 1920s about their true nature.

From photographic plates taken over many years van Maanen had measured proper motions in the spiral arms of some of the spiral nebulae, like

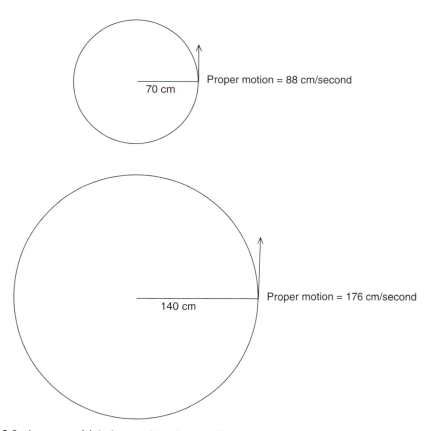

Fig. 5.3. A stone whirled round in the smaller circle has a lower speed compared with the stone whirled in the larger circle, provided they complete one circle in the same period.

M 33, M 81, M 101, etc. Proper motions are movements of celestial bodies
across the line of sight. To get the idea of a proper motion think of a rope with
a stone tied to it, being whirled round so that the stone makes one revolution
in, say, five seconds. The stone has a speed in space that is perpendicular to the
rope. Suppose the rope length is 70 centimetres. Then the stone describes
a circle of radius 70 cm. The circumference of the circle described by the stone
is therefore $2 \times \pi \times 70 = 440$ centimetres, approximately. (We have taken
the value of π to be 22 / 7 approximately.) Therefore we calculate that the stone
is moving at the speed of $440 \div 5 = 88$ centimetres per second. Now we make
the rope double its original length, that is, 140 cm. The stone is still whirled
round one revolution in five seconds. But its speed in space is now doubled to
176 cm per second.

These speeds of 88 cm/second or 176 cm/second are the proper motions of
the stone in the two cases as viewed from the centre of the circle. So we have the
simple rule that if the stone is kept whirling so as to make a round in a fixed
time, its proper motion will be larger, the farther it is from the centre. Notice
also that for an observer looking at the moving stone from the centre of the
circle, the stone will appear to change its direction.

Using this technique, van Maanen calculated the speeds at which the spiral
arms in the nebulae would have to move if they were at the enormous distances
found by Curtis and other observers. Van Maanen claimed to have measured
tiny changes of direction of the nebulae from year to year which enabled him
to estimate their speeds. These speeds, as we saw in the above example,
depended directly on the distance of these spiral nebulae. If they were at the
enormous distances suggested by Curtis the speeds could be of the order of
15 000 kilometres per second. Van Maanen then argued that if these nebulae
were similar to our Milky Way, they would be ripped apart if they were
spinning with such high speeds. So he gave a strong 'No' to the idea that these
nebulae were distant objects.

Curtis in his lecture in the debate with Shapley admitted that this objection
would hold, but he expressed caution in drawing such conclusions based on
observations that required very tiny changes of direction. Indeed, the kind of
changes of direction that a proper motion measured by van Maanen required
were as tiny as a *hundredth of a second of arc per year*! To those readers
unfamiliar with this measurement of angle we may add that the typical small
division on a protractor used for measuring an angle by school children is
a degree. Divide the degree into 3600 parts to get the tiny angle of one arc
second. The claim for observing shift in the direction in a nebula by van
Maanen was a *hundredth* of this small angle. Such small-scale changes, if
they are really present, are extremely difficult to measure.

Since van Maanen was a classical observer with a strong reputation for accuracy, which his other measurements in stellar astronomy enjoyed, this objection weighed very strongly with many contemporary astronomers. Indeed, this was the major obstacle towards the realization of the fact that the spiral nebulae were indeed very distant galaxies well beyond the confines of the Milky Way. How did this fact finally get acceptance?

Hubble's measurements of distances of galaxies

At the same time, there was shaping up another line of investigation of the types of nebulae themselves. The principal player in this development was Edwin P. Hubble. Growing up in Wheaton, Illinois, USA, in the first decade of the 1900s. Hubble was noted more for his exploits on the athletic field than in scholastic studies. He had set a state record for high jump in Illinois and was also a good boxer. However, he later became one of the Rhodes Scholars at Oxford University where he studied law. He had, however, also studied astronomy and mathematics at the University of Chicago, where he had taken his first (B.S.) degree. He took his Ph.D. at the University of Chicago in 1917 and after serving in World War I, in 1919 he took up a position at the Mount Wilson Observatory in Southern California at the invitation of its director George Ellery Hale.

The following decade of the 1920s was to prove very productive in Hubble's life. The superior observing conditions available at the Mount Wilson Observatory were put to good use by him in embarking upon what turned out to be the new field of extra-galactic astronomy. That is, of astronomy relating to the universe lying beyond our galaxy, the Milky Way.

We shall encounter Hubble's name in several different contexts. Here we first meet it as part of the ongoing controversy about the distances of several nebulae, especially spiral ones. How do we determine the distance of an astronomical object?

As far as stars are concerned, we have seen that the distances of nearby stars are determined by the method of parallax (*see* Chapter 3). The method involves looking at a star at an interval of six months and measuring its direction against the background of more remote stars. The gap of six months ensures that we are observing the stars from two locations maximally separated from each other as our moving platform of the Earth orbits the Sun. We have seen that the apparent direction of the star would change as our vantage point changes. However, we also noted that the change will be very small if the star is very far away. Indeed, the method becomes unreliable for stars further than, say, a few hundred light years.

Fig. 5.4. Edwin Hubble shown with the 60-inch telescope at the Mount Wilson Observatory. Photo by courtesy of the Mount Wilson and Las Campanas Observatories.

This method will not therefore work for stars further than a thousand light years; and certainly cannot be used to determine the distance of a spiral nebula if we suspect that it lies beyond the Milky Way, at a distance of more than several hundred thousand light years. It may be tempting to ask whether we cannot use the same trick as we did for parallax measurements, but now with the orbit of the Sun around the Galactic Centre. The answer is alas, no! For the Sun takes approximately 250 million years to make one orbit and we cannot wait for periods of the order of several million years to make two measurements. So we need some other method to measure the distance of a nebula.

We have seen that such a method exists and has been used for stars in the Galaxy. Although not as accurate as the parallax method for nearby stars, this method nevertheless does give us an accurate enough estimate of distance for our purpose, namely to decide whether the nebulae are part of the Milky Way or are galaxies in their own right lying far outside it. The method requires the identification of a sufficiently bright star with special identifying characteristics in a typical nebula.

Suppose we do succeed in this programme and find a star X in the nebula. That is, we are able to say that star X in the nebula is very similar to star Y in our Milky Way, so much so, that we can assert that both X and Y are of the same luminosity. That means that if X were placed by the side of Y, we would

Fig. 5.5. Henrietta Leavitt (1868–1921). Credit: *Pencil sketch by Anagha Pujari.*

find them equally bright. However, if, as we suspect, the nebula is very distant, then X will be also and so it will look very faint. The actual observed faintness of X will depend on its distance. We know from our experience with stellar distances, that if X is 1000 times as far away as Y, then it will appear to be 1000 × 1000, that is a million times fainter than Y. Using this argument in reverse, we see that if we can measure how much brighter (or fainter) the star X looks compared to Y, we can estimate its distance.

The argument works if we can identify a reliable class of stars as candidates for X and Y. This is what Hubble was able to do. He was fortunate in this respect because a class of stars known as *Cepheid variables* had been identified with clear characteristics just a few years earlier. This very important contribution had come in the year 1912 from Henrietta Leavitt at the Harvard College Observatory.

Henrietta Leavitt had joined the Observatory as a volunteer in 1895. She was appointed to the permanent staff in 1902, and eventually became Chief of the photometry department. She worked there for the rest of her life. In these days of automation and computer-reliance it is perhaps hard to imagine the intense effort required to collect and sift through the data. But, with such efforts Leavitt discovered 2400 variable stars, about half of the known total in her day. These discoveries led her to the study of a special class of stars very similar to the star *Delta Cephei* which had been known since 1784. These stars were therefore called *Cepheid variables*.

Delta Cephei was discovered by John Goodricke in 1784. Historically, it was the second such star to be discovered; the first, *Eta Aquilae*, had been discovered earlier the same year by Goodricke's good friend, neighbour and collaborator Edward Pigott. The special characteristic of these stars is that their brightness fluctuates up and down and in a regular way. For example, the star Delta Cephei varies with a period of 5 days 8 hours 37.5 minutes and during one cycle, it shows a quick and sharp rise from minimum to maximum, and slowly declines to its minimum again. The maximum luminosity of the star is approximately two and a quarter times its minimum luminosity. Today the star can be observed by amateur astronomers as a binary star.

Henrietta Leavitt especially studied stars that are variable in their light output. And those that looked like the original Delta Cephei she studied with special attention. She found many such stars in two nebulae called *Magellanic Clouds*. These nebulae are so named because they were first spotted by the pioneering Portuguese seafarer Magellan (*c.* 1480–1521) as luminous clouds and they are very conspicuous as one flies into the Southern Hemisphere. These are known today as satellite galaxies of our Milky Way. Henrietta Leavitt found that although these Cepheids had different periods of variation

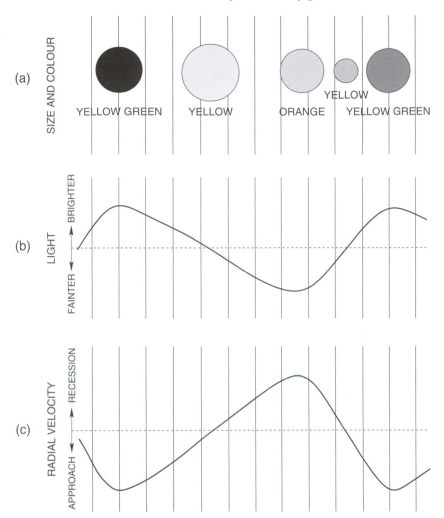

Fig. 5.6. The Cepheid variable star shows periodic changes in its size, colour and light emitted per second. It is seen to be pulsating by measurements of the speed of its surface coming towards the observer and going away from them. These changes are indicated against time. The changes typically take place over a period of a few days.

of brightness, they also differed in their maximum brightness. As they were all located in the same cloud, she rightly assumed that they are at nearly the same distance from us. By intense observation and mathematical calculation, Leavitt realized that there is a direct relationship between a star's brightness and period of variation. The brighter the star the longer it took to get through its cycle of maximum and minimum light.

This important finding is associated with Henrietta Leavitt. In science, there are natural relationships between measured physical quantities. Often these

The world of galaxies

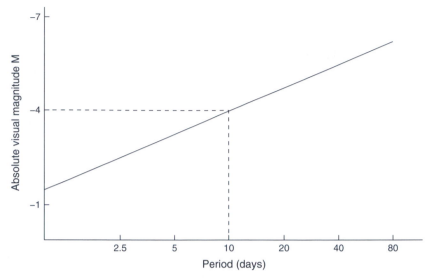

Fig. 5.7. The changes taking place in a Cepheid variable have a fixed period, measurable in days. The longer the period, the more powerful is the star in terms of the light it emits. This figure shows how the absolute brightness increases with period, a relation found by Henrietta Leavitt in 1912.

are not obvious as they get buried under errors of measurement. Nevertheless, by analyzing a large sample a careful observer can discover the hidden relation. The relationship discovered by Leavitt is known as the period–luminosity relation.

Leavitt first published her findings in 1912 in a chart of 25 Cepheid periods and their apparent brightness. From Cepheids of known distances, one can work out their luminosities. Thus we have a period–luminosity relation. From the period–luminosity relation one can read off the luminosity of any newly discovered Cepheid from its observed period. Using this relation therefore, astronomers only needed to know the period of a Cepheid variable to figure out how bright, and therefore how far away it was. Until then, methods for measuring distances in space only worked within about 100 light years. With Leavitt's findings, distances of Cepheids could be determined up to 10 million light years. This was the stepping stone for astronomers like Hubble to probe the extragalactic world.

Thus during the decade of the 1920s Hubble, and other astronomers were able to establish the distances of many spiral nebulae and to demonstrate that they are indeed galaxies in their own right at distances of many millions of light years. Hubble showed in 1924 using this method that the Andromeda Nebula is the nearest spiral galaxy to our own; we now put it at a distance of about two million light years.

Galaxy types

Apart from the very important work of establishing the extragalactic nature of spiral nebulae, Hubble went on to study the shapes of the nebulae and classify them accordingly. One can see examples of spiral nebulae in the following photographs.

The names given to these galaxies, beginning with the letter *M*, or the letters *NGC*, indicate the catalogue they are from. We have already mentioned the *Messier Catalogue*, which uses the letter *M*. The *New General Catalogue* uses the letters *NGC*. Henceforth we may refer to specific galaxies by their catalogue names as in the figures. Here one may see letters *Sa, Sb, Sc* in parentheses

Fig. 5.8. Spiral galaxy M 101. The spiral structure denotes concentration of the stars along the spiral arms. Credit: *Telescope In Education*.

Fig. 5.9. This galaxy (NGC 1365) has spiral arms coming out of a central barlike structure. It is called a barred-spiral. Credit: *European Southern Observatory*.

against the catalogue names. These indicate their 'class' as per Hubble's criteria. The typical galaxy of spiral class has a central nucleus and spiral arms. The *Sa* type has a dominant nucleus with very small and tightly bound spiral arms. As we go down the sequence, we have galaxies typically with a smaller nucleus and longer spreading arms. Our Galaxy and the Andromeda are both of type *Sb*.

The next class of galaxies is the *barred spirals*. These have a bar instead of a round nucleus and spiral arms like the ordinary spirals. An example of

Fig. 5.10. Elliptical galaxy M 32. Credit: *Kitt Peak National Observatory*.

a barred spiral is shown in the figure. These are labelled *SBa*, *SBb*, *SBc* in the same sequence as for spirals.

The spirals seem to dominate the population of galaxies. However there is another major type called the *elliptical* galaxies. Spirals may not dominate in the universe, but they do appear to be more common among the brighter, nearby galaxies which are readily seen. If we use larger telescopes which can see down to fainter nebulae, the majority component turns out to be made of the ellipticals. This is an example of what astronomers call a *selection effect*. One can understand this effect by an example from daily life. If one visits an airport and surveys the income of the population of air travellers found there, one may think on the basis of the data so acquired, that the population of the country is well-off. However, this may very well be an overestimate of the average income of the country as a whole, for, only the relatively well-off people can afford air travel.

The ellipticals have, as their name implies, an elliptical (egg-shaped) appearance and seem to be filled with older stars and relatively little gas and dust. The spirals seem to have young as well as old stars and have more gas and dust than the ellipticals. M 32, whose image is shown on the previous page, is an elliptical. The ellipticals are labelled in a sequence *E0, E1, . . ., E7* starting with those which are nearly spherical and going down to the increasingly flatter versions.

The overall sequence of galaxies is depicted in the 'tuning-fork'-like diagram below with the elliptical sequence branching into two spiral types, barred and unbarred. This diagram is due to Reynolds, an English astronomer, and Hubble. It would be wrong to believe that galaxies evolve from one type to

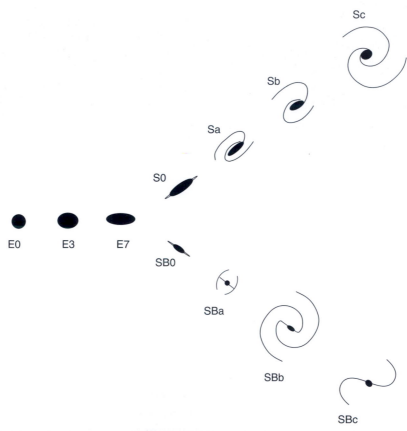

Fig. 5.11. This diagram, due to Edwin Hubble, places the sequence of galaxy shapes in a concise form and because of its shape, is often known as Hubble's tuning-fork diagram. It is, however, *not the case* that galaxies evolve in these sequential modes.

Fig. 5.12. This galaxy M 82 is neither a spiral nor an elliptical. It is called an irregular galaxy as its shape does not fit any set pattern. Credit: *Hubble Space Telescope image gallery.*

another along this sequence. Indeed, we do not yet know how galaxies are born and how they evolve. In later chapters of this book we will air some of the current views on the formation of galaxies.

This account will not be complete without mentioning that there are galaxies, which do not fit into this scheme of ellipticals and spirals. There are *peculiar* galaxies that show some exceptional feature, like a jet, for example, and *irregular* galaxies that show no simple features at all. One of these is shown in the image above. It may well be that these peculiar or irregular galaxies contain the clues to the formation of galaxies.

What we have described here is just about all that was known about galaxies by the end of the 1920s. Hubble had established the extragalactic nature of many of the nebulae. However, the difficulty with van Maanen's observations remained. They had not gone away, nor had they been retracted. Hubble did not believe them, but he diplomatically avoided any confrontation with his senior and experienced colleague in the same observatory. He did get another astronomer at Mount Wilson, Seth Nicholson to remeasure van Maanen's plates. Nicholson could not confirm van Maanen's results, but nothing was ever published.

Eventually the data were essentially forgotten (or ignored) and the stage was set for the most remarkable discovery about the universe, again to be made by Hubble.

6

The expanding universe

Why is the sky dark at night?

We interrupt our narrative concerning Hubble's work on galaxies to ask the question raised above. The reader may wonder why we raise this simple question about a local observation from Earth, in the midst of a discussion about far-away galaxies. The connection will soon become clear.

At first sight the answer seems simple enough. The Earth spins about its axis with a period of twenty-four hours and the part of its surface facing away from the Sun experiences darkness, which is nightfall. Is this not a satisfactory answer to the question?

Heinrich Olbers, a German physician and astronomer was not satisfied with this answer. In 1826 he carried out a simple calculation and arrived at an answer so startling, that it kept astronomers busy for a century and a half trying to find where Olbers had gone wrong. For if he were right, then his conclusion was that the sky should not have been dark at all, but extremely bright all of the time, irrespective of which side of the Earth the Sun was on. Known as the *Olbers paradox*, the argument used by Olbers is essentially as follows.

Besides the Sun, the sky contains many other stars which are also emitting light, some of which will reach the Earth. Of course, the amount of light from a typical star will be quite minuscule compared to what we receive from the Sun, because the star is very far away. However, Olbers argued that although the amount of light from distant stars will be very little there are so many stars in the universe, that their combined light might not be negligible. He set out to compute this using a simple argument.

Imagine that the universe is infinite in extent and is uniformly filled with stars, all like the Sun. Suppose we draw a sphere of radius R and consider a thin shell on its surface as shown in the diagram on the next page. Our geometry textbook tells us that the surface area of this sphere is $4 \times \pi \times R \times R$, a result we have already used in Chapter 3. Now, if the shell has thickness a, its volume will be this area multiplied by the thickness, that is, $4 \times \pi \times R \times R \times a$. Further, if the universe is uniformly filled with stars and there are N of them in a unit volume, then the total number of stars in the shell will be given by simply multiplying the shell volume by N, that is by $4 \times \pi \times R \times R \times a \times N$. Now imagine a typical star in this shell has a luminosity L. Then, as we discussed in Chapter 3, the amount of its radiation coming our way per unit area per second would be L divided by the factor $4 \times \pi \times R \times R$. So we see that by multiplying this quantity by the number of stars in our shell, these stars contribute a total flux of radiation equal to $L \times a \times N$. Notice that all terms relating to distance have cancelled out in our arithmetic, or, rather in the arithmetic used by Olbers. So we should get the same amount of light from a shell of the specified thickness a, *no matter how far away it is*.

The last part of Olbers' argument is now straightforward. Divide the entire universe into concentric spherical shells all of the same thickness. Each shell

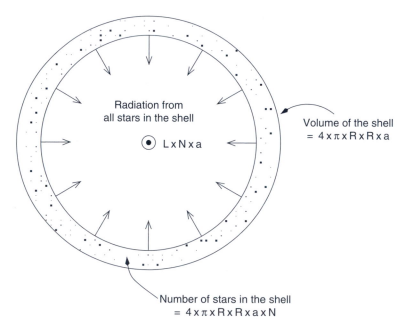

Fig. 6.1. A spherical shell drawn in space with the observer O at the centre. The shell has a specified thickness. The further it is from O, the larger its volume and the larger the number of stars in it. However, the light received from a faraway star is very feeble and so the combined light from all the stars in the shell is the same, whether the shell is far or near. (The feebleness of light per star is compensated for by the large number of stars in a remote shell.) Based on this effect, Olbers had argued that the total light received by O by all such shells far and near in an infinite universe would be infinite.

contributes the same flux to the observer. *But the number of such shells is obviously infinite*. Therefore it follows that the total flux from *all* stars in the universe is also infinite! This was the logical conclusion that Olbers came to with his rather simple basic assumptions. Thus it is immaterial whether we are facing the Sun or not. Either way the night sky will be infinitely bright. But, when we face away from the Sun, the night sky *is* dark. So there is something wrong with the arithmetic. But where is the mistake?

Careful consideration of all of Olbers' arguments shows one loop-hole. The stars are not point sources, they have a finite size. So when we start putting stars in successive shells around us there will come a point when they will block the entire sky visible to us. An analogy may help here. If you look through a gap between trees in a park you can see the buildings in the background. However, if you are in a thick forest of trees, you simply can't see beyond the foreground trees which block the view of trees further back. So, in the revised Olbers calculation, only the stars in the relatively nearby shells will contribute to the total radiation flux. The total flux is therefore not infinite but finite.

But we are not out of the woods yet! For this finite total flux can be computed and it still turns out to be very high, as high as on the surface of the Sun. This means that the sky should not only be bright but the temperature in our neighbourhood should be as high as it is on the surface of the Sun, in the region of about 5500 degrees Celsius. Again, we seem to have arrived at an impossible conclusion.

Astronomers suggested two other ways out of the Olbers paradox. The first is that the universe may not be infinite as Olbers assumed, but it is finite in extent. Which means that when we draw our spherical shells we stop at a certain distance beyond which nothing exists. This distance would have to be at least as large as the range of our best telescopes. For, as far as we can see, there is no end to the sources of light up to the distance of some ten billion light years that we can presently probe. If indeed there are no more sources of light beyond, say, ten billion light years, we do get a resolution of the paradox, since the contribution of sources out to this distance is negligible compared to the light we get from the Sun.

The other possible solution is that the stars that we see, or can in principle see, came into existence a finite time ago. Suppose the universe itself came into existence ten billion years ago. Then we can receive light only from those stars that lie within a distance of ten billion light years. Light from stars that exist beyond this limit, has not had time to reach us yet. The figure below illustrates this scenario.

Another possible resolution of the paradox takes note of the fact that stars in any shell will only last for a finite time. They cannot go on shining for ever,

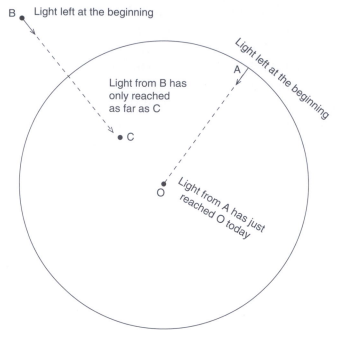

Fig. 6.2. If the universe was created, say, ten billion light years ago, light received today by observer O could not have come from a distance further than ten billion light years. That is, Olbers' calculation could not go on to infinite distances and would be limited to a finite value. In the figure above, the furthest distance is shown by the boundary sphere.

so we cannot expect to find shining stars in all shells. This also reduces the net contribution to the total flux received.

Thus we can see that what started as a simple question of local interest has forced us to think of cosmological issues such as the extent and age of our universe. A further crucial element that has been ignored, is the discovery by Edwin Hubble that the universe is expanding. We now turn to that piece of evidence which was not available to Olbers. It was the most important discovery made in cosmology in the last century.

The spectrum of a galaxy

One of the earliest discoveries concerning the properties of light was that by Isaac Newton more than 300 years ago. Newton held a glass prism against sunlight and observed that the light entering the prism came out split into several colours, very similar to the seven colours of a rainbow: violet, indigo, blue, green, yellow, orange and red, often grouped under the acronym

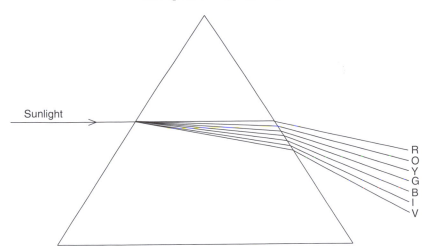

Fig. 6.3. When light enters a glass prism from an oblique angle at its surface, it is split into several colours, ranging from violet to red. This happens because light waves of different wavelengths are deflected by different amounts as they pass from one medium (air) into another (glass). Violet waves are deflected by a greater angle than red waves. This light is deflected again when emerging from the prism into air as shown above. This was the effect first noted by Newton.

VIBGYOR. As shown in the figure, light is bent by the prism and the violet colour is bent the most, red the least. Newton argued, rightly, that light of all these colours is present in sunlight. With its property of bending light of different colours by different amounts, the prism is able to act as a separator of these different components.

Today we know that light travels as a wave and the wavelength of each colour is different. The violet waves have the shortest wavelengths while the red waves have the longest. The entire range of wavelengths covering the seven colours is from about 400 nanometres to 800 nanometres. A nanometre is a very tiny unit of length; a billion nanometres make a metre. So the basic rule we arrive at from the prism experiment is that light of different wavelengths is bent differently, the bending being larger for the shorter waves.

Astronomers use this effect to split the light coming from a distant source like a star or a galaxy. They have built instruments called spectroscopes that produce a *spectrum* of colours, these colours being the components of the light from the source. The amount of energy being carried by the different bands depends on the physical mechanism that powers the source. A given source has a spectrum, and if we can interpret this spectrum we can learn something about the source.

The spectrum of the Sun obtained by Newton showed the typical rainbow colours. When we look at the same spectrum taken with a modern

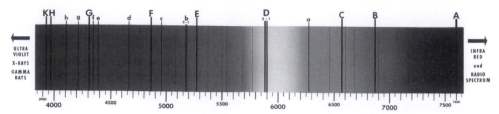

Fig. 6.4. The solar spectrum showing the Fraunhofer lines as dark lines indicating absorption of sunlight. *(Courtesy of the Indian Institute of Astrophysics, Bangalore, India.)*

spectroscope, we not only see the seven colours, but also some dark lines spread over it. Prima facie, dark lines denote absorption of light. Why is light absorbed selectively at a few wavelengths?

The dark lines were discovered by Joseph von Fraunhofer in 1814, and their origin was a mystery until the dawn of quantum theory. Quantum theory describes the physical behaviour of matter at the microscopic scale. The rules, that govern the motion of an electron inside an atom for example, are very different from those that govern the motion of a planet around the Sun. These rules show that nature on a very small scale does not behave in a smooth, continuous fashion, but shows discrete behaviour. We will not go into the details here, but simply relate this behaviour to the dark lines. The light coming from a star to us, passes through the outer layers of the star (its atmosphere) and the atoms in that outer part are cooler, and absorb some of the light. The absorption takes place, not smoothly over a band of colours, but over discrete sets of lines at specific wavelengths. These wavelengths are tuned to the specific structure of the absorbing atoms. Applying the principles of quantum theory, one can relate the wavelengths to those atoms.

These dark lines in the Sun's spectrum are called the Fraunhofer lines, and they are identifiable with the absorbing atoms of hydrogen, calcium, magnesium, iron, etc. Each line has a well-defined wavelength at which it occurs. The lines are called *absorption lines*.[1] The information they bring is of great help to the astronomer in many ways. For example, they tell us what material is present in the absorbing material, its temperature and also the motion of the source carrying it. This last named property is known as the *Doppler Effect*.

Discovered by Doppler, first for light waves, the effect is also valid for other waves such as sound. The sound effect is easier to identify in daily life. When a source of sound is approaching us, its pitch appears to increase: it sounds

---

[1] There are some bright lines too in the spectra of some sources. These are called *emission lines*. They arise from excited atoms which carry greater energy than normal atoms. These atoms emit, rather than absorb radiation at specific wavelengths and these emissions are seen as bright lines.

shriller. On the other hand, when the source is receding the sound appears flatter. One can experience this effect when standing on the platform of a railway station as an express train passes at great speed, its whistle blowing. The whistle sound appears shrill and turns to flat as the approaching train passes by and then recedes.

The change of pitch is in fact due to a change in wavelength of the sound waves. The increase in shrillness corresponds to a decrease in wavelength while the shift towards flatness is due to an increase in wavelength. A simple rule is that for an approaching train the fractional decrease in wavelength is equal to the ratio of the speed of the source to the speed of sound, and likewise for a receding train a similar result obtains for fractional increase in wavelength.

The Doppler effect can also be applied to light since that too travels as a wave. Following the analogy of sound waves, it therefore tells us that the shift of the dark lines increasing the wavelength or decreasing the wavelength will be related to the recession or approach of the source. Suppose, for example, that in a spectrum a dark line should appear at a wavelength of 500 nanometres. However, in the actual spectrum of a source it is seen at a wavelength of 550 nanometres. The increase is 10% and it equals the ratio of the speed of recession of the source to the speed of light. Knowing that light travels at a speed of 300 000 km per second, we conclude that the source is moving away from us at 10% of the speed of light, that is at 30 000 km per second.

With this technique, astronomers have studied the spectra of many stars beyond the Sun and have measured their speeds towards or away from us. Stars moving towards us have their spectral lines shifted towards the short wavelength side, that is the violet or blue end of the spectrum. For stars moving away from us these lines shift towards the longer wavelength side, that is, the red end of the spectrum. For this reason such shifts are referred to as *blueshift* and *redshift* respectively. Thus, quantitatively, a redshift z measures the fractional increase in wavelength. In the example of the previous paragraph, $z = 10\%$ or 0.1. From where we are situated in the Galaxy, we see some stars moving towards us and some moving away from us. These speeds are of the order of 100 kilometres per second or less. These are all parts of our own Galaxy – the Milky Way. But for external galaxies in general we encounter a very different situation.

Early in the last century, in around 1913 to 1914, Vesto Melvin Slipher at the Lowell Observatory in Northern Arizona found that the spectrum of the Andromeda Nebula shows a blueshift, corresponding to a speed of 300 kilometres per second towards us. This was a high speed by stellar standards. However, as Slipher continued to study the spectra of nearby spiral nebulae he found speeds of the same order; but gradually as measurements covered more red nebulae, the numbers of redshifted nebulae began to dominate. Slipher's 1925 paper reports that there were 11 redshifted nebulae to 4 blue-shifted ones and that by 1917, this ratio had increased to 21 to 4. In short, after a few blueshifted cases, the vast majority of nebulae showed redshifts.

Hubble's law

While he was working on the properties of the nebulae at Mount Wilson, Hubble was aware of Slipher's results. Hubble, however, looked at the images of the sources along with their spectra. Typically he saw a pattern emerge which we can understand with the help of a series of photographs of such sources side by side with their spectra as shown on the next page.

Notice that the images shown on the left become smaller and smaller as we go down the list. The clarity of the image and its brightness are also reduced as we go down the list. Prima facie we can argue that as we proceed down the list, we are looking at more and more remote sources of light. (Imagine that we are looking at a townscape from the top of a high tower, with houses dotted around at different distances: as we look further and further away, the houses look smaller and smaller and are less sharply defined.) On the right side we have the spectra of these sources with a dark line which has shifted towards the red end. The shift is larger and larger as we go down the list. If we appeal to the Doppler effect we can argue that, the further down the list we go, we are finding more remote galaxies and ones which are moving away from us at greater speeds. In short, speed goes with distance.

What Hubble had deduced from this work was that the velocity of recession – that is, the redshift – was proportional to the distance inferred from the apparent brightness. The further away the galaxies were, the faster they were moving. With his younger colleague, Milton Humason, Hubble made a systematic study of a large number of galaxies and clusters of galaxies. In 1929 Hubble published this result in the *Proceedings of the National Academy of Sciences*. We show his graph on page 84.

Notice that Hubble has plotted the distance of the galaxy and its radial velocity away from us. The trend is clear, bearing out what the above examples of images and spectra indicated. *The speed of a galaxy away from us is in proportion to its distance from us. This is known as Hubble's law.* So if we have two galaxies *A* and *B*, with *A* twice as far away as *B*, then we should find that *A* is moving away from us with a speed twice that of *B*.

This seems at first sight to indicate that Hubble's observation has finally restored to us a privileged status in the universe! We appear to be located at

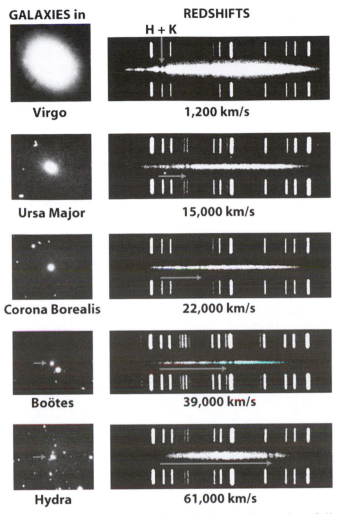

GALAXIES in

Virgo

Ursa Major

Corona Borealis

Boötes

Hydra

REDSHIFTS

H + K

1,200 km/s

15,000 km/s

22,000 km/s

39,000 km/s

61,000 km/s

Fig. 6.5. This photograph has several galaxies in increasing order of distance as one goes down the list. The left-hand column has the image of a galaxy while the right-hand column has its spectrum. Hubble found that as one goes down the list, the images get smaller and fainter, as one would expect from increasing distance. Likewise the spectra show increasing redshifts as shown by the arrows. Credit: *Palomar Observatory and California Institute of Technology.*

a vantage point from where all other galaxies are moving away in a highly symmetrical manner (*see* Figure 6.7). Alas this is not so! In fact the Hubble law further strengthens the highly democratic set up of the universe. For, if we imagine transporting ourselves to another galaxy (any one of those we see moving away from us) we can work out and convince ourselves that we would see exactly the same Hubble law operating from our new vantage point as the centre of the universe.

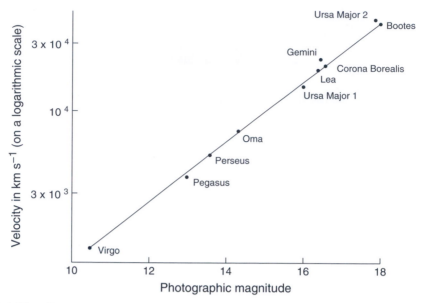

Fig. 6.6. The diagram representing Hubble's findings on the velocity–distance relation, in 1929. The horizontal axis represents the apparent magnitude of a galaxy or cluster. The apparent magnitude is a measure of the distance of a light source: the further away a source is, the fainter is its appearance and therefore the larger its apparent magnitude. The vertical axis represents the radial velocity of the galaxy away from us. Notice that the points in the diagram show that the radial velocity is higher for the more distant galaxies.

The analogy of a balloon being blown up will help understanding of the situation better. Imagine, as shown in Figure 6.8, we have a balloon on which are stuck tiny discs of paper indicating nebulae. As we blow up the balloon, the discs all move away from one another. An observer on any of these discs will see other discs receding as per Hubble's law. In reality no disc on the balloon occupies a 'central' position.

Drawing on the same analogy we may argue that in reality it is the universe that is expanding like a balloon being inflated. Indeed this is the conclusion that the astronomers were confronted with: *that we are all living in an expanding universe.*

Although this was the interpretation that most astronomers subscribed to, there was an alternative explanation advocated by the astronomer Fritz Zwicky in 1929. Zwicky suggested that light particles, the photons, lose energy through absorption, scattering etc. in the intergalactic space and as a result their wavelength increases. This explanation is known as the 'tired light hypothesis'. This has been proposed from time to time to understand the redshifts of galaxies and

(i)

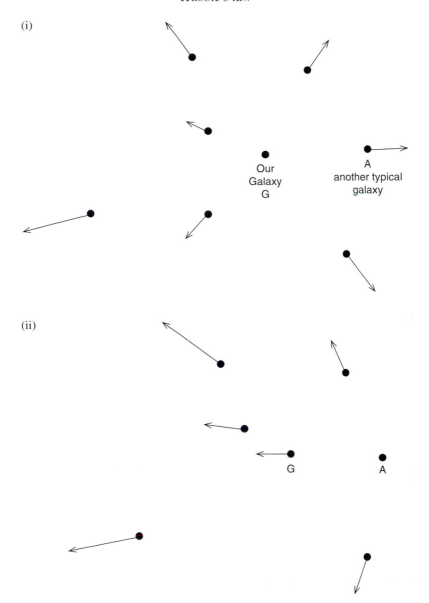

(ii)

Fig. 6.7. (i) Galaxies viewed from our Galaxy G are seen to be moving away as per Hubble's law. Thus a neighbouring galaxy A is moving away from G with a speed in proportion to its distance from G. (ii) What will an observer on A see? To that observer the galaxy G will appear to move away in the opposite direction and likewise he or she will see all other galaxies also moving away radially as per Hubble's law. In short, all observers are alike so far as the Hubble law is concerned. Man on this Galaxy need not consider himself as occupying a specially privileged position in the universe.

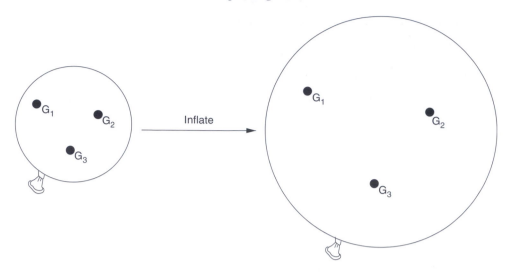

Fig. 6.8. The balloon analogy for the expansion of the universe.

quasars; but by and large it has very few adherents because it is in conflict with the laws of atomic physics.

Concluding remarks

The picture that was beginning to emerge as the decade of the 1920s drew to a close was that we are living in a dynamic universe in which the galaxies are moving in a highly symmetric fashion, and they are all receding from one another. If Hubble's law is universally applicable, it provides the astronomer with a powerful tool for measuring the distance of a remote source of light. Thus we can conclude that:

I. If two galaxies are close neighbours then their spectra will exhibit approximately the same redshift.
II. If two galaxies have different redshifts, then the one with the larger redshift is further away than the other.

We will work with these premises for the time being. Only when we reach Chapter 16 will we mention some of the modern findings that suggest that the above two conclusions may not be valid for all extragalactic sources.

We may at this stage return to the Olbers paradox. Recall that Olbers did his calculation for a static universe. What difference will an expanding universe make to the Olbers calculation? A major difference is that if the universe is expanding, then the light from distant galaxies will be highly redshifted. The quantum theory of light tells us that light is not only a wave, it is also a collection of tiny packets of energy called *photons*. The energy of a typical photon is determined by the wavelength of light. The longer the wavelength, the smaller is the energy. The light coming from a far-away galaxy will therefore be reduced in energy by the redshift effect. So the contribution to the energy from a remote source of light will be much less than that estimated by Olbers. In the calculation described earlier, we should find that light from a more remote shell of sources will be much less than the light from a nearby shell of the same thickness. It was pointed out by Hermann Bondi in the 1950s that once this effect is taken into account in computing the total energy received from the whole universe, we get a sensible answer. The contribution from the rest of the universe is negligible compared to the light we get from the Sun.

Modern observational cosmology is believed by many to have begun with Hubble's law. However, theoreticians will argue that they got into the game twelve years earlier! Why? We will find out in the next chapter.

7

Modelling the universe

From models to reality

The figure shown on the following page is that of a globe representing the Earth. It is a ball revolving around an axis which is inclined to the vertical direction at an angle of approximately 23 ½ degrees. This is the angle the axis of the real Earth makes with the direction perpendicular to the plane in which it moves around the Sun. The schematic diagram illustrates this. The globe thus gives us a means of visualizing reality: it is a model of the actual Earth, in the same way that a toy car is a model of the real car, giving a child some idea of what the real thing is like.

Physicists believe in making models to simulate the real object they want to study. Unlike the globe or the toy car, their models are abstract mathematical stuctures which can be useful in making us understand how real things behave. So, to understand what the Sun is like, we start by imagining that it is a huge ball of cold gas. How huge? That is decided by its total mass. A typical adult human being has a mass of around 50–70 kilograms. The Sun's mass is estimated to be about

2 000 000 000 000 000 000 000 000 000 000 kilograms!

Also, astronomical measurements tell us that the diameter of the Sun is 1 400 000 kilometres.

So our gas ball has to be that massive and very large. Using Newton's law of gravitation we can ask what will happen to this ball? Won't it shrink because all parts of it attract one another? It will, and a mathematician will tell us that with this force, the ball will shrink to a point in a very short time, namely less than twenty nine minutes. Clearly such a model will not do to describe the real Sun which has a fixed shape. So we need to be more clever in making a model. We get a better model if the gas is not cold but hot. For a hot gas has large pressure and this will help prevent the gravitational contraction of the ball.

Fig. 7.1. The globe shown above is a model of the actual Earth, used to understand its broad geographical structure. In the same way a model of the universe is a simplified version that helps us understand what the actual universe is like.

A physicist can then get into greater mathematical detail to work out how the pressure will have to change as we go inwards towards the centre of the Sun.

This line of reasoning can, of course, be continued further to get into a more detailed model of the Sun. However, we chose this example only to demonstrate how a physicist goes about trying to understand nature. Our main interest here is not in the Sun but in the universe as a whole. Can we make a model of the *entire universe* which will simulate what the astronomers are finding? Can we, for example, think of a model that describes an *expanding universe*, which Hubble's observations revealed?

The subject that deals with the large-scale properties of the universe including the expansion, is called *cosmology*, and a model of the universe is called a *cosmological model*. The purpose of a cosmological model is to help us to understand the large-scale behaviour of the universe. For example, we would like to know if the universe has been expanding in the past and whether it will continue to do so in the future. We would like to know how it came to acquire galaxies of the various types that we see today. And so on and so forth. . .To describe this exercise, we will begin at the beginning of such scientific attempts to model the universe. And that beginning takes us right back to Isaac Newton.

Newton's universe

In the 1690s, Isaac Newton attempted an ambitious application of his law of gravitation. He wanted to describe, with the help of his theory of gravity, the largest physical system that can be imagined – the universe. How did Newton fare in this attempt? Newton started by constructing a mathematical model of the real universe by making a few simple assumptions. He assumed that the universe is homogeneous, isotropic and static.

What do these adjectives mean? By 'homogeneous' one implies that all parts of the universe are having the same physical properties. Thus there is *no variation from point to point*. Likewise, the property of 'isotropy' implies that there is no variation from one direction to another when observed from any vantage point in the universe. In short, if you are blindfolded and taken to some place in this universe, on removing your blindfold you can tell neither where you are, nor, in what direction you are looking. And, 'static' of course means that the universe does not exhibit any large-scale motion of its component parts. However, Newton found that even such a simple model had some problems.

In a letter to Richard Bentley a divinity student, dated 10 December 1692, Newton expressed his difficulties in the following words:

It seems to me, that if the matter of our Sun and Planets and all ye matter in the Universe was evenly scattered throughout all the heavens, and every particle had an innate gravity towards all the rest and the whole space throughout which this matter was scattered was but finite: the matter on ye outside of this space would by its gravity tend towards all ye matter on the inside and by consequence fall down to ye middle of the whole space and there compose one great spherical mass. But if the matter was evenly diffused through an infinite space, it would never convene into one mass.

The figure shown, of a finite and uniform distribution of matter in the form of a sphere initially at rest, helps explain Newton's difficulty. Will such a sphere stay at rest forever? The matter in the sphere has its own force of gravity, which tends to pull the different parts of the sphere towards one another, with the result that the sphere as a whole contracts. Such a force of self-gravity is known to exist in stars, for example, because they are massive balls of matter. However, in their case the internal pressures present in the gases that make up stars oppose self-gravity and maintain the stars in a static shape. What happens to the matter distribution in Newton's universe?

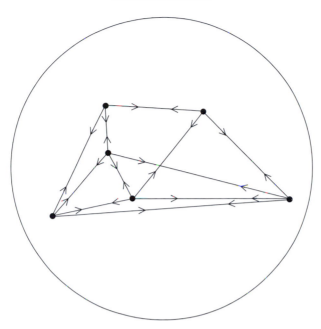

Fig. 7.2. In a sphere there are forces of gravitational attraction between any two points, pulling them closer towards each other. The result of all these forces is to make the entire sphere contract.

Pressure force acts between regions of low and high pressure. In the case of a star, the pressure close to the centre is higher than that in the periphery. Because of our assumption of homogeneity, the pressure forces do not operate in Newton's universe at all! So the sphere undergoes continuous contraction: *it cannot remain static*. Not only would the sphere shrink, it would also attract any bits of matter located outside, as Newton noted in his letter.

There is a difference, however, between our *finite* system of the above figure and an *infinite* universe. The finite sphere collapses toward its centre. Where should the infinite universe collapse to? A little consideration will show that there is *no central position* in a uniform infinite universe. So there is no *net* tendency to collapse!

To put it differently, consider a typical point *P* in our universe. If we take into account the overall pull of attraction of all points in the universe, we find that *P* is pulled with equal intensity in all directions with the result that pulls in opposite directions cancel out and *P* stays where it is!

Although from this theoretical reasoning a static universe is possible, Newton realized that it would be highly *unstable*. A small departure from uniformity would lead to an *enhancement* of that departure. For example, if a certain part of the universe has slightly larger density than the average, then that part will

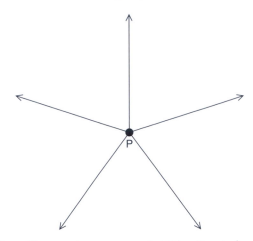

Fig. 7.3. In an infinite uniform universe any point P is attracted equally in all directions by the gravity of matter. So there is no resultant force on P in any direction, so the point stays where it is. By this argument Newton demonstrated that the entire universe would remain static.

preferentially pull matter towards itself and so will *grow the density of matter in its neighbourhood*. As a result, the universe would break up into finite-sized clumps of matter that would undergo gravitational collapse. Newton realized this problem before long and in his next letter to Bentley, dated 17 January 1693, mentions this instability:

And much harder it is to suppose that all ye particles in an infinite space should be so accurately poised one among another as to stand still in a perfect equilibrium. For I reckon this as hard as to make not one needle only but an infinite number of them (so many as there are particles in an infinite space) stand accurately poised upon their points.

It appears that Newton abandoned his attempts to model the universe any further.

The Einstein universe

In 1915, Albert Einstein formulated a new theory of gravitation, which was more comprehensive than Newton's law of gravitation. Einstein viewed the phenomenon of gravitational attraction between any two chunks of matter as a property of the space in which they are located and the time measurements that are made regarding them. A simple example will illustrate the difference between Einstein's and Newton's points of view.

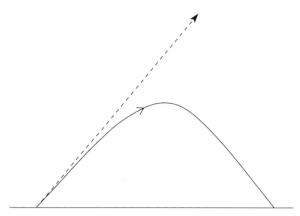

Fig. 7.4. A projectile fired into the air at an angle to the vertical would continue to move in a straight line (the dotted line in the figure) with uniform speed, if there were no force acting on it. In reality, there is a force acting on the projectile, that of the Earth's gravity. So the projectile follows the curved (parabolic) trajectory shown by the continuous curve. This was Newton's way of looking at this situation. Einstein argued differently! To him, the situation (represented by the dotted line) of no gravity has no *locus standi* in the real world. Only the continuous trajectory is real and it represents uniform motion in a straight line, but in a space not following Euclid's geometry. Thus the continuous line actually followed by the projectile is 'straight' according to the new geometry that applies to the space permeated by Earth's gravity.

Imagine a projectile thrown at an angle to the vertical as shown in the figure. We know that it traces a curved trajectory, rising as it moves away from the point of projection and eventually falling to the ground. This curved track is the result of the gravitational force of the Earth on the projectile. Had this force not been there the projectile would have continued to move in a straight line, going up and up. For this is what Newton's first law of motion states: a body continues to move in a straight line with uniform speed if there is no force acting on it.

However, Einstein's treatment of this problem would be different. He would argue first that it is meaningless to talk about the situation of the second kind since there is no way we can check what the projectile would have done *had there been no gravity of the Earth acting on it*. Gravity is not an effect that can be removed or wished away. If it is present in space, it will be there as a permanent entity. *In short, it is as permanent as space and time in that region.* Taking his cue from this argument Einstein assumed that gravitation is an effect that is intrinsic to space and time. And in a master-stroke of genius he argued that the way to relate gravity to space and time is through the medium of geometry.

Geometry is a subject we have encountered at school. It deals with points and straight lines, with angles and triangles, rectangles and circles and so on.

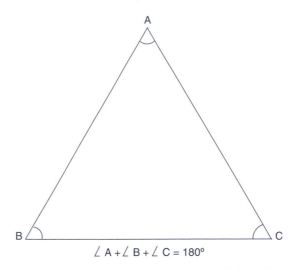

$$\angle A + \angle B + \angle C = 180°$$

Fig. 7.5. Any triangle drawn on a plain piece of paper will have its three angles adding up to 180 degrees. This is what we learn and prove in Euclid's geometry taught in school.

The subject tells us about relationships between different measurements through its many theorems. For example, we have come across the theorem illustrated in the above figure.

This theorem tells us that the three angles of a triangle add up to two right angles or, 180 degrees. However, we also recall that theorems in geometry are proved on the basis of certain basic assumptions, called *axioms* and *postulates*, by the originator of the subject, Euclid some 23 centuries ago. Geometry based on Euclid's assumptions is called Euclidean geometry.

What Einstein proposed was that *Euclid's geometry does not apply to measurements made in space and time where and when gravitation is present*. Indeed long before Einstein came on the scene, mathematicians had found several such 'non-Euclidean' geometries that are as logically consistent as Euclid's geometry. An example will suffice to illustrate this circumstance.

Imagine that we need a geometry that will help us measure distances and angles on the Earth. The Earth is not, however, flat...it is curved like a ball. The globe with the map of the Earth on it that we encountered earlier is a good model of the Earth. Suppose we station three observers on the Earth, one at the North Pole and two on the Equator, at the points of its intersection with the zero degree (Greenwich) meridian and the 90 degree meridian. Figure 7.6 illustrates this situation. Let us call the three observers *A*, *B* and *C*.

Suppose that a triangle is drawn between vertices A, B and C. Thus straight lines connect *A* to *B*, *B* to *C* and *C* to *A*. They do so following the definition of a straight line: it is the path of shortest distance between the two end-points.

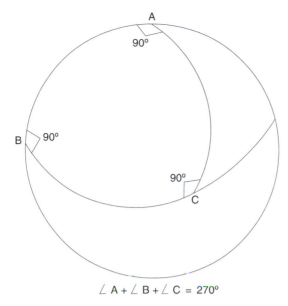

Fig. 7.6. On the surface of the Earth, however, the theorem illustrated in the last diagram does not hold. The triangle drawn on the Earth as shown above, has a right angle at each of its three vertices. So the three angles add up to 270 degrees!

To connect *A* to *B* such a track will lie along the rope that is stretched between these points and lying on the Earth. One can easily check with the help of the globe, for example, that the straight line *AB* is the meridian of zero degrees. Likewise, the straight line connecting *B* to *C* on the Earth is the stretch of the Equator linking them and *C* to *A* is the 90-degree meridian between *C* and *A*.

Now comes the strange conclusion that *A*, *B* and *C* will arrive at: strange because they have been brought up (like all of us!) on Euclid's geometry. The angle between the straight lines *AB* and *AC* as measured by *A* will be 90 degrees or one right angle. So will be the angles at *B* and *C*. So the triangle *ABC* has three angles adding up to *three right angles*. We thus have a contradiction with our earlier Euclidean theorem which stated that the three angles should add up to *two* right angles.

At this stage you may protest that this example involves cheating! You may argue that the lines *AB*, *BC*, *AC* are curved: they are not straight. However, you have to remember that we are living on the Earth and on its spherical surface these are the lines of shortest distance and hence straight lines.

Taking his cue from such examples, Einstein argued that whenever gravitation is present space and time will have curvature like the Earth and so the geometry governing measurements of distances and angles will not be Euclidean. Going back to our example of the projectile, Einstein would argue that in

Fig. 7.7. Albert Einstein (1879–1955).
American Institute of Physics: Emilio Segrè Visual Archives.

the geometry on the surface of the Earth, as modified by the Earth's gravity, the trajectory followed by the projectile represents motion with uniform speed in a straight line. The trajectory will of course not appear straight to those who insist on using Euclid's geometry; but if one uses the correct geometry one would discover that the projectile is indeed moving in a straight line with uniform speed.

How does one discover the 'correct' geometry? Here Einstein came up with a set of equations that make up his general theory of relativity. These equations tell us how to determine the geometry if the physical contents of space and time are known. Whatever gravitational interaction exists between these objects will be described by how these objects move in such a space, and this in turn can be explained once one knows the geometry of space and time.

This was the Einstein programme that sought to replace Newton's law of gravitation. Tests made on planets and other objects in the Solar System bear out Einstein's expectation that the results show general relativity as giving a more accurate description of nature than Newton's theory.

In 1917, barely two years after formulating his theory of gravity, Albert Einstein applied the theory to the ambitious task of constructing a model for the universe. Like Newton, Einstein also thought that the universe is static on the large scale. Like Newton's attempted model of the universe, Einstein's universe was also imagined to be a homogeneous and isotropic distribution of matter. However, when Einstein tried to apply his equations of general relativity to this model, he too encountered a fundamental problem. To put it bluntly: *his equations did not have a solution that corresponded to his model.*

Faced with this difficulty, there were two options open to Einstein. Option 1 was to abandon the model and look for a different one that came out of his equations. Option 2 was to modify his equations so that he could get his desired model from them. He took the second option. He found that if he added a new force to the equations he could get his model. This force predictably was one that opposed gravity (recall that for want of such a force Newton could not get a static spherical model), and its strength was measured by a constant usually denoted by λ or Λ. This constant is called the *cosmological constant*, and it was destined to play a mixed role in cosmology, as we will see later.

Nevertheless, with this constant Einstein was able to obtain a static model of the universe that was homogeneous and isotropic. It, moreover, had a geometry different from Euclid's and according to the model the universe was *finite but unbounded*. This may appear to be a contradiction in terms. How can something be finite in extent and yet have no boundary? A simple example of such a structure in one dimension is that of a circle. One can go round the circle but not come across an end anywhere. Thus it has no boundary. Yet its length (the circumference) is finite. Similarly the surface of a sphere (like the Earth) is finite in area but has no boundary. These are instances from our world of experience. We need to stretch the imagination one notch higher and imagine a three-dimensional space that is the 'surface' of a four-dimensional hyper-sphere. That is the space that Einstein's universe was modelled after.

This result pleased Einstein as this model not only fulfilled all the simple properties demanded of it, but it also showed how non-Euclidean geometry can play a role on the largest scale. It demonstrated the basic hypothesis of general relativity that, given sufficient matter, it can curl up the space into an unbounded but finite volume. He believed that this was a unique solution provided by his modified equations. In a way this also demonstrated to him that his programme of modification was along the right lines.

Enter de Sitter

This satisfaction was not to last for long. Within a few months the Dutch theorist William de Sitter came up with another solution of Einstein's modified equations. He found a model of a universe that was empty but expanding! If it was empty, then what exactly was expanding? You may well ask. The answer is that if one identified points in space with galaxies, but assumed that these galaxies had no mass to attract one another, then these will be moving away from one another.

Of course, in 1917 astronomers believed that the universe was static and so to most of them, this kind of model was at best a crazy mathematical solution with no physical reality to describe. The astronomer Arthur Eddington suggested that the Einstein model describes matter without motion and the de Sitter one, motion without matter. Nevertheless, to Einstein, the de Sitter solution came as something of a shock. While he still believed in his own static model, its non-uniqueness, especially in the form of a de Sitter-type solution was certainly unwelcome.

In retrospect, we see the de Sitter universe as having some merit too. For, it describes an expanding universe, which Edwin Hubble's observations were to establish in 1929. However, in 1922, another physicist entered the limited field of modelling the universe. His impact, though not seen at the time, was to be lasting on the development of mathematical cosmology.

The expanding universe

Alexander Friedmann was a young Russian physicist who decided in 1922 to look at Einstein's original equations, that is those without the cosmological constant, to see if he could get *non-static* models of the universe. That is, Friedmann assumed that his model would describe a universe that is homogeneous and isotropic but not at rest. At the time there was no great observational motivation to look for such a model. As we saw in Chapter 6, observations of spectral shifts of nebulae were beginning to be made, but there was no appreciation amongst the astronomers that these nebulae were moving away from one another.

When Friedmann dropped the assumption of a static universe, he discovered that he could get solutions of Einstein's equations. In particular, he could get solutions in which the universe expands. That is, the distance between any two galaxies steadily increases with time. One can visualize a Friedmann

Fig. 7.8. Alexander Friedmann (1888–1925).
Leningrad Physico-Technical Institute, courtesy AIP Emilio Segrè Visual Archives.

model as a cubical grid of metal wires which are heated. When a metal wire is heated it expands, and so the intersections of the grid wires will move away from one another as shown in Figure 7.9.

An alternative way to understand this model is to go back to the balloon analogy given in Chapter 6, where the spots on the expanding balloon are seen to recede from one another.

Our everyday notion of expansion may fail us in fully understanding what is going on here. For the balloon analogy makes us ask, *what is the universe expanding into?* The feeling behind such a question is that just as the balloon

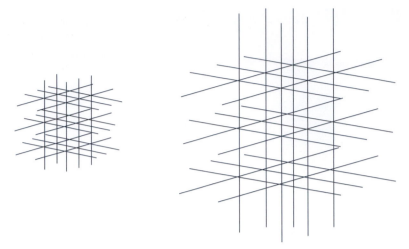

Fig. 7.9. This grid of metallic wires would expand if heated, making all its vertices recede from one another. The model universe of Friedmann was like this with galaxies as vertices. . .and as Hubble found, the galaxies do seem to be receding from one another.

is perceived to be expanding in ambient space, the universe must also be expanding in some space containing it. *That is not quite correct!* For, the universe is the whole space. It is not part of something bigger containing it. Rather, a better way to understand the notion is to imagine that the space in the universe has all points marked and as we observe it, these points all move away from one another. This is where we feel that the expanding grid analogy is more representative of the fact.

The further success of the Friedmann model is that if we calculate the velocity of recession of any of these points from one another, we find that this velocity increases in proportion to the distance between the points, just as Hubble's law found. So with today's hindsight we can say that the Friedmann model describes the actual observations of the 'velocity–distance relation' of galaxies made by astronomers like Hubble and Humason.

'Hindsight' is right; for when Friedmann sent his mathematical paper to Einstein for comments, he did not get any encouraging pat on the shoulder from the great man. For Einstein at the time did not see the relevance of the Friedmann model to reality. In 1922 astronomers still believed in a static universe and so Einstein could not be blamed for treating the Friedmann models as the kind of abstract curiosities that mathematicians delight in. History is silent about what response, if any, Friedmann received from Einstein.

In 1924, Friedmann obtained additional solutions by using Einstein's modified equations with the cosmological constant inserted. These naturally included the special solutions that Einstein and de Sitter had obtained. Most of these

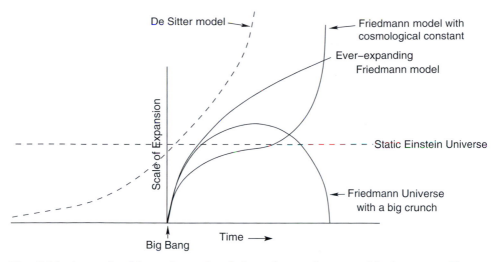

Fig. 7.10. A graph of how the scale of the universe changes with time according to different Friedmann models. We may interpret the scale as typical distance between two galaxies. Notice that most models start with zero scale factor, that is a state when all matter in the universe was compressed in a point, that is in zero volume. This state came to be identified with the *big bang*, according to the nomenclature given by Fred Hoyle. A typical model starts its expansion from the big bang. As we see, in their later behaviour different models differ from one another. Some expand and then collapse, some expand but with slower and slower rates while some accelerate at a later stage.

models describe the universe as emerging from a point-like state with an enormous explosion and then continuing to expand in all directions. In some cases the expansion slows down and the universe comes to a temporary 'halt', that is, a static state, and then begins to contract. Like a movie of expansion shown backwards, this contraction accelerates and eventually the whole universe is crushed to a point with a state of infinite density. If the beginning was in a 'big bang' the end is in a 'big crunch'. In other Friedmann models the expansion continues for ever. If there is no cosmological constant, the expansion slows down as time goes on, although it never comes to a halt. In such a case the universe would diffuse out into a state of zero density. If there is a cosmological constant, it eventually shows its effect by accelerating the expansion. Again, such a universe would diffuse out to a state of zero density.

Friedmann died in 1925 and did not live to see posterity recognize his work. Meanwhile, in 1927, Abbé Lemaître, a Jesuit priest who liked mathematics and was among the few who then understood Einstein's complex equations of relativity, solved them in the same way as Friedmann had done five years earlier. This was work done independently, for Lemaître was unaware of Friedmann's work. However, he was aware of the growing set of observations of redshifts of galaxies and saw that the expanding models provided a way of

Fig. 7.11. Abbé G. Lemaître (1894–1966). Credit: *Archives Lemaître. Université catholique de Louvain. Institut d'Astronomie et de Géophysique G. Lemaître. Louvain-la-Neuve. Belgique.*

understanding them. He also saw that his mathematical model predicted a 'velocity–distance relation'. He worked out the constant of proportionality that Hubble was to arrive at from his observations two years later.

Lemaître worked in the Institut d'Astrophysique at Liège, and his 1927 paper is proudly displayed there as a demonstration of his perception of the phenomenon of the expanding universe, earlier than Edwin Hubble. As we will see in this book again and again, there are many instances where due recognition is not given to the origin and originator of a novel idea or to the discoverer of a new fact, because it happened ahead of times when the scientific world was not ready for it, or simply because of lack of proper communication channels that publicize new facts and ideas amongst scientists.

Lemaître, unlike Friedmann, lived long enough to see his model receive recognition amongst cosmologists. As regards the origin of the universe, he saw that the initial point-like state suggested a sufficiently dramatic epoch for it. He called the compact state of the universe just after 'creation' a *primeval atom*. Start the universe at this stage, he argued. What went on earlier is not within the scope of science. We will return to this way of thinking later.

Basic theoretical principles

The theoretical models used by Friedmann and Lemaître follow certain basic symmetry rules which make the job of the model builder simple. Mathematician Hermann Weyl and theoreticians H. P. Robertson and A. G. Walker have given a rather precise form to these assumptions. These assumptions have influenced cosmological model builders since those early days. We will now briefly describe them.

When we look at an assembly of moving objects we can have a variety of situations. At one end we have an army of soldiers marching in a steady column. The soldiers march in such a way that they preserve their formation with no collision or disruption, with their steps keeping pace. At the opposite extreme is a mob in civil unrest. . .with individuals on the rampage in a random fashion. In the former case, it is very easy to describe the state of the individual in the group. . .where he is and at what speed and in what direction he is marching. In the second case, this is clearly not possible.

Weyl's postulate states that in the universe the large-scale motion of matter is more like the first case above. If we assume that most matter is confined to galaxies, we can equate galaxies with the soldiers and say that the galaxies move in a systematic fashion along well-defined trajectories, never colliding and fanning out, in such a way that we can state clearly at what time which galaxy was where in space. This allows the model builder to imagine a universal clock that all galaxies follow. The time measured by this clock is the *cosmic time*. Just as at a given time in the march of the army we can state that all soldiers have their right foot forward, so can we make assertions about the state of the universe as a whole at a given cosmic time.

And how does the state of the universe look at a given cosmic time? This is stated by the so-called *cosmological principle* (CP). The CP states that at any given cosmic epoch the universe looks homogeneous and isotropic. That is, from a study of the distribution of galaxies in the universe, one cannot make out any privileged position or any specific direction. Note that Newton, Einstein, de Sitter, Friedmann and Lemaître were using these assumptions without explicitly stating them. In the 1930s, Robertson and Walker introduced them into cosmology in a formal mathematical framework.

At this stage let us recall Aristotle's penchant for circles mentioned in Chapter 2. He liked them for their symmetry, which is no different from the symmetry envisaged in the cosmological principle just described.

Open and closed universe

The advantage of a mathematical model is that it permits us to make specific predictions. Some of these can be tested by observation. If observations bear them out, well and good; we continue to put our trust in the model. If the predictions fail, then we have to go back to the drawing board. Either the original model is superficially wrong and can be fixed by cosmetic surgery, or we have to abandon it altogether and look for a new model. This scientific method can be applied to the simplest Friedmann models, that is to those *without the cosmological constant*, in the following way.

We mentioned that although all models seem to begin in a point-like state, the dynamical behaviour of the universe is not unique. We can be more specific now. For this we introduce the notion of a 'critical density'.

Consider an astronaut walking on the Moon. He finds that it is much easier to make a high jump on the Moon than on Earth. This is because the force of gravity on the Moon is about a sixth of that on Earth. If the astronaut weighs 60 kg on Earth, his weight on the Moon will only be around 10 kg. Clearly it is much easier for him to lift 10 kg up in a high jump than 60 kg. The muscular force that generated the jump will carry the astronaut much higher on the Moon than on Earth because the gravitational pull is much less on the Moon.

Now consider an expanding universe. According to the Friedmann model, it exploded with a big bang. It starts moving outwards with the force of the primeval explosion. However, it is not entirely free to move outwards: there is the force of gravity on all its matter pulling it in. *Thus the expansion will be slowed down.* By how much? That will clearly depend on how much matter is stuffed into the universe. The greater the density of matter the stronger the pull of gravity and the stronger the tendency to slow down. So the universe decelerates after the big bang, although it will continue to expand. And as it expands it thins out and its density decreases. Will it slow down to a halt before it has thinned out so much that its density has become zero? If this happens then the universe will begin to shrink and end up in a big crunch.

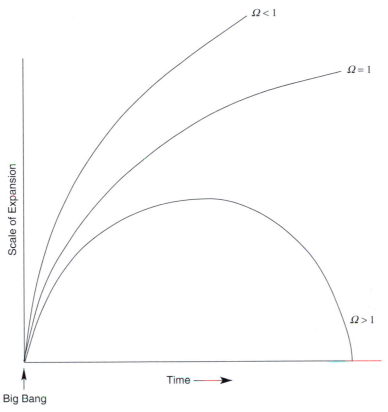

Fig. 7.12. This figure separates out the different Friedmann models according to their density. Those which have a density greater than the critical density subsequently contract. Those whose density is less than the critical density will expand for ever. The parameter Ω measures the ratio of the actual density to the critical density.

If the gravitational pull is just enough to slow down the universal expansion to a halt by the time the density of the universe has become zero, the matter density corresponding to that situation is called the critical density. Clearly, if the matter were more densely packed than this critical value, then we would expect the universe to slow down to a halt, contract and end in a big crunch. On the other hand if the matter density were less than the critical value then the expansion will never come to a halt and the universe will 'fizzle out' to a state of zero density. The figure illustrates the three cases.

There is one symbolic way of expressing this result. We can in principle measure the ratio of the density of matter in the real universe and the critical density and denote it by the symbol Ω. If Ω is greater than 1, then we have greater than critical density and we conclude that the universe is closed. If, on the other hand Ω is less than 1, we live in an open universe. The case $\Omega = 1$ describes a 'flat' universe.

There is another aspect of the critical density that relates to the geometry of space. Consider the three-dimensional space at any given time in a Friedmann universe. Will it have Euclid's geometry? One way to find out is to draw a huge triangle in space and measure its three angles. They should add up to 180 degrees if the geometry is that of Euclid. The Friedmann model predicts that if the density is critical, then this will be so. On the other hand if the actual density of matter exceeds the critical value, then according to the Friedmann model, the geometry will be *non-Euclidean* in such a way that the triangle in space will have its three angles adding up to a value greater than 180 degrees. Likewise the other case of non-Euclidean geometry arises when the density is less than critical density and the three angles of our triangle will add up to a sum less than 180 degrees.

In the former case, we say that the space is curved *positively*, and is of finite volume but without a boundary. Recall that the space in Einstein's universe was precisely of this kind. In the latter case we say that the space is *negatively* curved. It is of infinite extent in all directions and has no boundary. Another way to name these spaces is to call them *closed* and *open* respectively. The intermediate case of Euclidean geometry then gets the name: *space of zero curvature* or *flat space*. In this context, cosmologists refer to the critical density as the *closure density*. If the actual density exceeds closure density then the universe must be of the closed kind.

What is reality? Is the universe open or closed? How do we find out? We will discuss this very interesting issue in the following chapter where we will begin with Hubble's own efforts to come to grips with these questions.

8

What is the geometry of the universe like?

Hubble's quest

Having achieved the landmark discovery that the universe is expanding, Hubble devoted himself to a very ambitious task, a task that was to occupy him during the latter part of his life and which was to remain incomplete. In the 1930s he put the then biggest telescope, the 100-inch (2.5-metre) Hooker telescope at Mount Wilson, to the task of the wholesale observation of galaxies with the following purpose.

In the previous chapter we have seen that once we adopt Einstein's general theory of relativity for model building, then the simplest models of the universe hold out three options: (1) the universe is open, (2) the universe is closed and (3) the universe is flat. We saw that we can, in principle, decide which option is applicable to our universe by drawing a large triangle in space and measuring its three interior angles. If they add up to *exactly* 180 degrees then we live in a *flat* universe. If they add to a sum *greater* than 180 degrees then we live in a *closed* universe and likewise if the sum turns out to be *less* than this value then we are in an *open* universe.

Drawing triangles in space extending to several million light years is not a practical proposition and so the astronomer has to look for another test to settle this issue. A test which might be feasible is suggested by the following analogy in two dimensions.

Imagine you try to cover a flat surface, like a table-top with a sheet of paper. You place the paper on the top and try to smooth it over with your palm. You find that the sheet lies *flat* on the table top. Try to repeat the experiment with the surface of a sphere. You find that the paper gets wrinkled as your palm tries to press it all over the surface. This happens because the surface of the sheet has a *greater* area than that required to cover the surface of the sphere. On the other hand if we take our sheet to a saddle and try to cover the curved surface of the saddle, our paper will be torn, indicating that the area of the covering sheet is *less* than that needed to cover the surface of the saddle.

Hubble used this criterion with regard to *volumes* rather than to areas, since the open or closed nature of the universe related to volumes measured in space. To see how this can be applied to space imagine a sphere drawn in space with radius R. Euclid's geometry tells us that the volume of this sphere is $4 \times \pi \times R \times R \times R \div 3$. Here π is the constant we encountered before as the ratio of the circumference to the diameter of a circle: its approximate value is the fraction 22/7. However, if we draw a sphere in a *closed* space, its volume will be *less* than this value and likewise, if we draw it in an open space, the corresponding volume will be *more*.

Now imagine that you have drawn such a large sphere in space and you wish to measure its volume. One practical way will be to count how many galaxies it has. The more the galaxies the greater the volume. For example, in the case of flat (Euclidean) space if we double the radius of our sphere the volume will grow eight times. (In our formula above if R is doubled, the product is multiplied by $2 \times 2 \times 2$.) However, in the case of a sphere drawn in closed space the increased volume will not be eight times the original volume but will have grown less. Similarly for a sphere drawn in an open space the growth will be larger than a factor of 8. So if galaxies are filling the space uniformly (as is assumed in most models) then the way the number of galaxies grows as we count them to greater and greater distances will tell us what type of geometry applies to our universe.

This is the programme that Hubble undertook on the advice of the Caltech theoretician R. C. Tolman. By counting galaxies to successively larger distances Hubble hoped to find out whether their number increased more slowly or more rapidly than the number that a Euclidean model would predict. On the face of it the experiment seems simple. Going by our example above, if, in such a survey there are 100 galaxies up to a distance of say, 100 million light years, then in the flat Euclidean universe we should expect 800 galaxies out to 200 million light years. If we find only 600 up to that distance, then we live in a closed universe and if the number were as high as, say, 900 then we live in an open universe.

Reality is not so simple, as Hubble was to discover to his cost. There are several problems facing such an exercise. We enumerate a few of them.

The curvature of space is expected to be very small and to confirm this, one needs to go to quite a large distance. Distances of the order of a few hundred million light years are not adequate. In the diagram shown we see three typical curves A, B, C corresponding to the flat / closed / open universes. They begin to differ significantly only beyond around 3000 million light years. These distances were far greater than the distances that Hubble had measured so far.

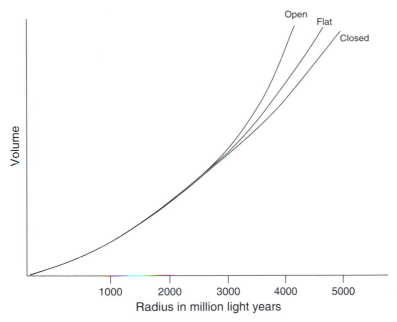

Fig. 8.1. In principle, one could count galaxies in universes of different types. If we carry out the exercise in an open universe, we expect the numbers to rise *faster* than in a flat Euclidean universe. Likewise, in a closed universe, the rise will be less steep. The different curves in the figure illustrate this expectation.

An analogy will help clarify this point further. We know that the Earth we live on is curved like a spherical ball. However, in a purely local context, like travelling within a city, we are not conscious of this fact. All our houses and roads are built as if we are on a flat surface. Travelling on the high seas where visibility extends to several miles, we are able to detect the effect of curvature. A steamer far enough away but approaching us, is invisible until it begins to appear above the horizon. Thus the difference between a flat Earth and a curved Earth can be detected only by covering a large enough distance. The same criterion appears when we try to measure the curvature of the universe.

The galaxies start becoming very faint as one tries to observe them at greater and greater distances. There are no reliable ways of measuring their distances. And, even if one accepts Hubble's law on trust for large distances, the

Fig. 8.2. The example of the approaching steamer on the high seas as it becomes gradually visible illustrates the curvature of the Earth and the horizon.

measurement of redshifts of such distant galaxies demanded more superior technology than that possessed by Hubble at that time.

The number of galaxies to be counted runs to millions rather than to hundreds and the technique of counting galaxies to very faint limits was no mean task. Only in the late 1980s did astronomers acquire automatic computerized methods of counting so many galaxies. So technology was not ready to provide tools for Hubble's test of counting galaxies.

However, Hubble used the inadequacy of the Hooker telescope to settle the cosmological issue, to press for a bigger and better telescope. The 200-inch (5-metre) telescope on the Palomar Mountain was the outcome of these efforts. The telescope, completed in 1948 was named the Hale Telescope after the astronomer George Ellery Hale. For many years it held the record for being the world's largest working telescope.

An interesting anecdote was narrated by Nobel Laureate astrophysicist Subrahmanyan Chandrasekhar about this telescope. When the proposal of building this telescope was approved, there was a press conference at which Hubble as well as Eddington were present. The press men asked: 'Sir, what do you expect to find with this telescope?' To this question the scientists replied: 'If we knew the answer, there would be no need to build this instrument.'

This repartee may be judged against the background of present proposals to build telescopes and other scientific instruments. To get the proposal to move

Fig. 8.3. The Hale Telescope with a primary mirror of 5-metre (200-inch) diameter. For nearly half a century this was the largest working telescope in the world. *Courtesy of Kshitija Deshpande, private collection.*

forward, the concerned investigating scientists have to outline clearly *what they expect to find with the telescope*. The proposal is accepted if the expectation is considered reasonable. An open-ended reply like that given by Hubble and Eddington today would get the proposal nowhere. And when the telescope is made and put to use the investigators have to justify by results that their expectation was correct. This approach to science funding makes it difficult to discover any unexpected results, results of the kind that have provided new directions for future growth.

The advent of radio astronomy

The Hale telescope was completed while Hubble was still firmly in the saddle as the world's leading observer in cosmology. However, by the end of the 1940s, the subject had advanced far enough to suggest to most astronomers that the programme attempted by Hubble of determining if the universe is open or closed, by counting galaxies, was not workable. The problems mentioned above were still very relevant even for a bigger telescope like the Hale telescope. Even if he never openly admitted that this was the case, Hubble gradually lost interest in that programme and turned to other extragalactic observations. The 200-inch telescope was turned to make other important observations, some of which we will describe at a later stage.

Fig. 8.4. Karl Jansky standing in front of his antenna.
National Radio Astronomy Observatory.

Fig. 8.5. Grote Reber with the antenna fabricated by him in his back garden. Credit: *Patricia Smiley, NRAO.*

In the meantime, astronomy was undergoing renaissance in the form of a new window of observations opening out on to the sky. The new branch, radio astronomy, was initiated in the 1930s by the pioneering efforts of Karl Jansky. While working for the Bell Telephone Labs on a project to investigate why there was interference in the ship to shore transmission of radio messages, Jansky had set up a large antenna. With its help he found that there were radio waves coming from outer space, interfering with the local transmission. As the intensity was rising and falling daily he thought that waves from the Sun were interfering with the network. This suggested a discovery of astronomical significance in the sense that the Sun was not only radiating light but also radio waves, and it naturally led to questions like whether other stars also radiated radio waves. However, it turned out that although the Sun is a weak source of radio waves, it was the Milky Way galaxy whose radiation was the culprit. Jansky, for example, was able to show that there was radio emission coming from the direction in which the centre of the Milky Way lay.

In the early 1940s, Grote Reber, an engineer by profession, set up his big antenna in the back yard of his house in Wheaton, Illinois, and made measurements of radio waves coming from different parts of the Galaxy including the galactic centre. It is said that when he sent a research paper on his findings for publication in the leading American research periodical, the *Astrophysical Journal*, the techniques being so new, the referee of the paper came to Reber's house to examine the set up. The apocryphal story is that he was asked to wait until Reber's mother cleared the antenna that she was using for drying clothes!

We remind the reader that in Chapter 6 we mentioned that light travels in the form of waves and that the rays that our eyes are sensitive to, have wavelengths between 400 to 800 nanometres (one nanometre is a billionth part of a metre). By contrast radio waves have wavelengths ranging from a few centimetres to several metres. But otherwise both visible light and radio waves are examples of the so-called 'electromagnetic wave'. When such a wave travels in space there are periodic ups and downs of electric and magnetic intensity just as the surface of water heaves up and down when a water wave travels through it. It was Heinrich Hertz who, during the period 1887–1889 demonstrated how

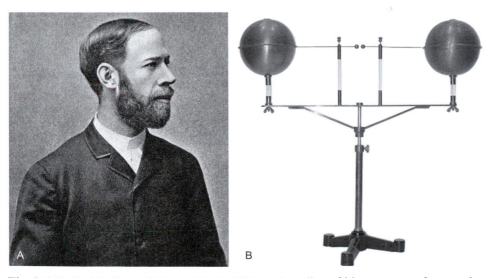

Fig. 8.6. Heinrich Rudolf Hertz (1857–1894) and replica of his apparatus for proving the existence of electromagnetic waves. National Radio Astronomy Observatory ("Astronomy" by Fred Hoyle, Crescent Books, 1962).

these waves can be produced and detected in the laboratory and how they travel through space. These are the waves which carry radio programmes to our wireless sets and which tell the radar detectors where the enemy aircraft is flying.

Cosmic radio sources

Much of the post-war radio astronomy owes its origin to scientists who had worked on radar during World War II. Bernard Lovell at Manchester University and Martin Ryle at Cambridge University in England, Jan Oort in the Netherlands and Bernard Mills at Sydney University, Taffy Bowen and John Bolton at Parkes in Australia were some of the early players in this emerging field. Lovell was responsible for setting up the large steerable dish of 250-feet diameter at Jodrell Bank near Manchester with a grant from the Nuffield Foundation, while Ryle built interferometers at Lord's Bridge near Cambridge under the patronage of Mullards. In Australia Mills built a gigantic cross-shaped antenna while efforts at Parkes led to the Australia Telescope National Facility.

The early efforts were to detect individual strong sources of radio emission, that is emission rising in intensity well above the background noise. Several radio sources were discovered this way including the source at the galactic centre found earlier by Jansky and Reber. For example a strong source of radio emission lay in the direction of the Cygnus constellation. Following the tradition of optical astronomers for naming sources after the constellation in which they are found, with Greek letters α, β, γ, etc., coming after in decreasing order of intensity, the source was named Cygnus A. These radio sources gave rise to a new controversy amongst astronomers reminiscent of the arguments at the beginning of the twentieth century about the location of the nebulae.

Tommy Gold, a maverick young scientist at Cambridge made the suggestion that most of these radio sources lay outside the Galaxy. His argument for reasoning this way was the observation that these sources seemed distributed in the sky in an isotropic way. . .that is, their distribution did not seem related to the shape and orientation of the Milky Way. This idea went contrary to the thinking of Ryle and his group in the Cavendish Laboratory, who believed that most of these sources were radio stars well within our Galaxy. The debate steadily grew more acrimonious to the extent that at one stage Gold was made to feel unwelcome at the Cavendish seminars on radio astronomy!

It took some years to resolve the controversy and the source Cygnus A played a leading role in the process. Using the Palomar 200-inch telescope in 1952, Walter Baade and Rudolf Minkowski were able to show that optically the source could be identified with a galaxy whose estimated distance (by today's yardstick) of about 600 million light years, put it well outside our Milky Way. This was the beginning of the evidence that ultimately vindicated Gold. Although this put an end to the distance controversy, it did nothing to cool the ruffled tempers on both sides. We will see an aftermath of this feud in the events to follow later.

The Cygnus A example demonstrated a weakness of radio astronomy vis-à-vis optical astronomy. The latter, in the form of Hubble's law gives a rough and ready measure of the distance of an extragalactic object. If we have the optical spectrum of the object, from which its redshift can be measured, then Hubble's law tells us that the distance of the object is approximately equal to the redshift multiplied by the speed of light divided by Hubble's constant. Thus a galaxy with redshift 0.01, say, gives the distance as 0.01 × 300 000 kilometres per second divided by Hubble's constant. Taking Hubble's constant as 20 kilometres per second per million light years, we get the answer as 150 million light years.

How does a radio astronomer measure distances? While there are lines in the radio spectrum they are not strong enough to be detected in very distant galaxies. Thus he cannot get a radio spectrum with strong spectral lines as his optical counterpart can. So he is not able to measure the spectral shift and hence cannot apply Hubble's law. The only way he can proceed is to appeal to his optical colleague to survey the part of the sky where the radio source lies and to identify which optical source is identified with the radio source. To do so, the radio astronomer has to give the position of his radio source very accurately so that the optical astronomer can look that much more precisely. In general the

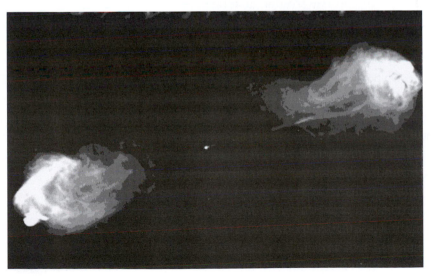

Fig. 8.7. The radio source Cygnus A. Credit: *Chandra X-ray Observatory*.

optical sky is densely populated and unless one narrows down the area of the sky in which the source lies, the exercise of optical identification is dicey! This is like searching for a needle in a large hall…unless one can narrow down the search area, the chance of success is small.

Walter Baade was given an 'error rectangle' small enough to attempt looking for the optical counterpart of Cygnus A and he came up with an optical source which was remarkable in itself.

At first sight the picture suggested two objects in close juxtaposition, and not surprisingly it was assumed to be depicting a pair of galaxies in collision. Baade himself strongly believed in this interpretation and for a few years the popular interpretation of an extragalactic radio source was that it represented a collision of galaxies. In this connection there is the story of a bet between him and the astronomer Rudolf Minkowski. The bet concerned whether the Cygnus A source really represented a collision of galaxies. Minkowski was sceptical and asked for a bet. Baade proposed that a test be made to examine the issue further. This involved taking a detailed spectrum and looking for what are called emission lines. These lines arise when atoms in the light source are in an excited state (that is, their electrons have higher energy than in the normal stable state). The excited electrons in these atoms jump down the energy ladder and in this process they emit radiation of specific frequency. This radiation appears as a bright spectral line. (Contrast this situation with the *dark absorption* lines mentioned in Chapter 6, which arise from absorption of specific radiation.) Baade and Minkowski believed (rightly) that the presence of emission lines indicated a dynamical and energetic disturbance of the gases which could arise from a galactic collision.

But what was at stake for the winner of the bet? Baade was so sure of being right that he proposed that the loser should give to the winner a crate of whisky. Minkowski was not so sure; besides he had just invested capital in a house. So he talked Baade down to a single bottle of whisky. This turned out to be a prudent step, for a few days later Minkowski turned up with a bottle of whisky for Baade. He had conceded the bet, for he had strong emission lines in the spectrum of Cygnus A. Nevertheless, to Baade's chagrin, as they sat discussing the evidence Minkowski consumed most of the bottle he had brought.

Do colliding galaxies produce radio sources?

In retrospect, Minkowski was actually justified in drinking the whisky he had brought. Moreover he himself was entitled to a bottle from Baade, for further evidence in the 1960s revealed that a photograph like the one of Cygnus A does

Fig. 8.8. A laboratory synchrotron machine to accelerate subatomic particles: this is the Super Proton Synchrotron tunnel at CERN, with part of the machine shown. Credit: *CERN*.

not show colliding galaxies. Rather the photograph shows a single galaxy which appears in two parts because it is divided by dust lanes. We have already encountered the misleading nature of interstellar dust (*see* Chapter 4). By obscuring the background stars the foreground dust creates the impression that there is nothing in between. Closer examination of the Cygnus A spectrum shows that there is *outward* motion of gas particles rather than inward motion as in a collision. The whole picture is thus indicative of an explosion at the centre of the galaxy. Figure 8.7 shows a modern picture of this radio source.

Another argument from theory also came up in 1958 to deal a blow to the collision idea. Given the shape and size of a typical radio source and its radio spectrum one can make several important deductions. One is that the shape of the spectrum suggests that the radiation is coming from fast-moving electrically charged particles, mostly electrons that are being accelerated by a magnetic field. In the laboratory, physicists had successfully constructed accelerators using the same principle, namely bending the trajectories of fast-moving electrons so as to make them move in circles, using a magnetic field. These are called 'synchrotrons'. And for the same reason the radiation in the cosmic sources came to be known as *synchrotron radiation*.

Using details of the synchrotron process, and the observed parameters of a typical radio source, Geoffrey Burbidge was able to calculate what is the least amount of energy such a source needs to have in order to be able to radiate at the observed rate. The answer he arrived at was startlingly high ... something like ten million billion billion billion billion billion joules. Or to put it differently, as much energy as the Sun would radiate in a billion billion years. How could such a vast energy reservoir be generated in the radio source? More disturbingly for the collision hypothesis, Burbidge pointed out that this requirement exceeded by a hundredfold the energy that could be generated in a typical collision of galaxies.

We will return to this issue of what empowers a radio source later. We will now consider the issue of whether or not the radio sources can be used instead of galaxies to find the geometry of the universe.

Radio astronomy and cosmology

The failure of Hubble's mission was due to the fact that he had too many galaxies to count and also that the galaxies visible with his telescope did not extend far enough to enable him to distinguish between the different possible geometries of the universe. What if we count radio sources instead of galaxies? Radio sources are not as abundant in the universe as galaxies and so far as counting goes they do not present the same level of difficulty as galaxies. Further, the radio astronomers believed that the faintest of the radio sources that their telescopes were able to pick up were further away than the faintest galaxies that the best of optical telescopes could observe.

So here was a chance for the newly emerging science of radio astronomy to make a significant contribution to cosmology. Could the radio measurements decide the issue of whether we live in an open or closed universe? Before we come to this interesting question we will describe another important development in cosmological theory that was developed in 1948, namely the proposal of the steady-state theory. This theory arrived on the scene just as the observational techniques available to the astronomer were getting into a higher gear of sophistication. The steady-state theory provided them with an ample opportunity to have a go at testing its predictions besides testing those of the Friedmann–Lemaître cosmology.

9

A universe without a beginning and without an end

A radical triplet

During World War II much of the research into ways of detecting enemy aircraft and submarines was conducted by civilians working for the British Admiralty. This work brought together many very talented mathematicians and physicists who would not otherwise have ever got together. The particular trio who made such an impact on astronomy and cosmology were Fred Hoyle (1915–2001), Hermann Bondi (1919–2005), and Tommy Gold (1920–2004). They were all educated at Cambridge University. Fred Hoyle was a Yorkshire-man who graduated with the highest honours in theoretical physics and turned to astrophysics just before the war. Hermann Bondi and Tommy Gold both came from Austria and their families had left Germany because of Hitler. Bondi was an extremely talented mathematician who graduated from Cambridge during the war and Gold had also graduated in physics during the war. Both Hoyle and Gold in different ways had intuitive and very creative approaches. After they came back to Cambridge in the late 1940s they thought and wrote about many astrophysical problems which had remained unsolved.

In this book we only discuss two of the major fields they worked on, namely cosmology – the steady-state universe – and the origin of the chemical elements.

The steady-state universe

Fred Hoyle was extremely excited at Hubble's discovery of the expansion of the universe, and according to Tommy Gold when they were meeting in Cambridge after the war, he never tired of talking about it and wondering what it meant.

Soon after Hubble's discovery more than twenty years before, it was argued that if the universe is continuously expanding it must have been much smaller

Fig. 9.1. The originators of the steady-state theory, Thomas Gold, Hermann Bondi and Fred Hoyle. Credit: *Fred Hoyle, private collection.*

and denser in the past, and if you take that argument to the limit, there must have been a time very long ago when all of the matter and energy in the universe were contained in an exceedingly small volume. Thus the universe must have come into being in a very special event and then exploded outwards.

As we have written elsewhere, this type of model of the universe had been predicted from Einstein's equations, first by a Russian meteorologist A. Friedmann, who died in 1925 before the observational discovery had been made, and then by Georges Lemaître, a brilliant scientist who was also a Jesuit priest in Belgium in 1927. He made the discovery independently of Friedmann, but brought it to the attention of Sir Arthur Eddington in Cambridge and others in the late 1920s. As is often the case in science this work had also been seen and understood by R. C. Tolman at Caltech in the US at about the same time and while we don't know this for a fact, it is very likely that Tolman at Caltech had described this work to Edwin Hubble.

Thus by about 1930, there was general acceptance of the universal expansion and of the view that the universe *began* as a very tiny condensed body. This beginning was entitled 'The Primeval Atom' by Georges Lemaître in a paper that he wrote in 1936. However at that time there was little understanding of the physics of what might have been going on when the universe was in this very condensed state.

However, the idea of a beginning and the evolution of the universe as it expands was attractive to many people, not at the least because there are similarities to the ideas of creation represented in the Bible, and in Western culture generally.

Hoyle, Bondi and Gold discussed the problem extensively, and finally Tommy Gold made them think creatively about the possibility that the expansion as it was currently observed, always remained the same and never changed. They had recently seen a movie entitled, 'The Dead of Night', which had a plot which evolved to an end (near the end of the picture) which was exactly the same as the beginning.

Gold provoked Bondi and Hoyle to see if they could find a solution to the cosmological equations allowing the universe to remain always the same – to be in a *steady state*. They found that this could be done, so in 1948 two scientific papers were published, one by Bondi and Gold basing the steady-state cosmology on what they called a *perfect cosmological principle*, and the other by Hoyle who developed a new field theory modifying Einstein's theory when the gravitational field is very strong. This is how the steady-state theory was born.

For the observed universe to remain always the same, matter must be created at a rate determined by the expansion, and the evolved older matter disappears over the horizon. But this idea is no more extravagant than the approach in which everything was created in the beginning in a gigantic explosion erupting from a point.

In the late 1940s Fred Hoyle gave a series of radio lectures on the BBC. The broadcasts covered astronomy and cosmology and he published this in a book called *The Nature of the Universe*, in 1950. This was very widely read. In his lectures Hoyle gave the name 'The Big Bang' to describe the conventional cosmology, and this name stuck though Hoyle was using it in a slightly derogatory way.

In the 1990s the magazine, *Sky and Telescope*, invited its readers to come up with alternative names for the standard cosmology, names that they felt were more appropriate to describe the theory. Several names were suggested, but when they were put to a popularity poll along with the existing name, 'The big bang' won the poll by a handsome margin!

For more details we begin with 'The big bang'. It denotes an epoch when the physical properties of the universe were indescribable...not only that, mathematically all the parameters specifying the physical state of the universe become either infinite (e.g., density, temperature) or zero (e.g., volume). One cannot talk of a 'before' in relation to this epoch. Thus to ask, how all the contents of the universe came into existence at this epoch or just after it, is a meaningless exercise. The sacrosanct law of conservation of matter and energy (meaning that matter and energy together cannot be created or destroyed; they can only be interchanged) broke down at this epoch. Not surprisingly scientists refer to the big-bang epoch as a singular epoch, i.e., one they cannot understand.

Many scientists are not disturbed by the fact that such an epoch appears in their theory. To them this epoch suggests an event that goes beyond scientific enquiry. To some of them, those brought up in a culture in which the prevailing religion suggests a divine act of creation, this is the meeting point between science and religion. Looked at dispassionately, applying the same yardstick that scientists apply to their theories in other branches of science, the appearance of such a singularity is an indication that the theoretical framework describing the universe is faulty. Hoyle, Bondi and Gold all shared an atheistic outlook in their upbringing and so did not have a cultural hang-up in questioning the reality of the big bang.

There was more direct and factually uncomfortable evidence that questioned the status of the big bang as a cosmic epoch of 'beginning'. This may well be described by the apocryphal story involving two Cambridge professors, Sir Ernest (later Lord) Rutherford and Sir Arthur Eddington. Rutherford was a nuclear physicist and had done pioneering work on radioactivity. He did not have a great deal of respect for astronomy and cosmology which he considered to be highly speculative subjects. Eddington, as we have seen was a leading astronomer and cosmologist. Meeting Eddington one day in Cambridge, Rutherford asked: 'Professor Eddington, what do you consider to be the age of our universe?' Feeling that he could floor this down-to-earth physicist with an astronomically high number, Eddington replied: 'Professor Rutherford, I estimate the age to be two thousand million years.' Whereupon,

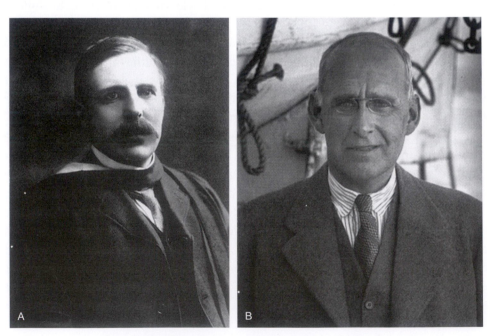

Fig. 9.2. Lord Rutherford and Sir Arthur Eddington.
Department of Physics, Texas A&M University – Commerce.

Rutherford produced a piece of stone from his pocket and said: 'Then it will interest you to know that I have just estimated the age of this rock to be three thousand million years.'

The technique used by Rutherford was that used by geologists to estimate the age of rocks. A typical rock has atomic nuclei of two kinds: stable or unstable. The stable ones retain their form for ever while the unstable ones change over the course of time with neutrons within them changing to protons. The neutrons are electrically neutral whereas the protons are positively charged. The decay process is known as radioactivity. The time scale for a specific unstable nucleus to decay in this fashion depends on its composition (i.e., on how many neutrons and protons it has). Some decay times are short, running into a few minutes while some are very long, ranging up to hundreds of millions of years. It is the latter that provide us with clocks for dating long ages like those of rocks. Thus the age quoted by Rutherford was typical of what geologists were finding, namely that the age of the Earth was as much as four thousand six hundred million years. . .much too large to fit into Eddington's universe of two thousand million years. In short, if the big bang did happen, and the universe came into existence then, that is, about two thousand million years ago, then we should not find any object in the universe older than this. As time has passed, the age of the universe has been estimated to be much larger than 2 billion years. But the ages of the oldest stars have from time to time appeared to be greater. We shall return to this issue in Chapter 11.

Then there was another aspect of the big bang that especially bothered Bondi and Gold. If the universe was indeed of the big-bang type then it was much hotter and much denser close to the big bang. For example, its density could be as high as a billion billion times the density of water and its temperature would run into a billion billion degrees. How do we ensure that the physical laws operating under such extreme conditions were the same as they are now? If not, how do we study such a universe scientifically? To give an example from another branch of astronomy, there are stars that are so compact that as much as a solar mass might be confined to a sphere of no more than a few tens of kilometres' diameter. To understand these stars, we must know the laws of physics that govern the matter they are made of. It turns out that this matter is made largely of neutrons, and its high density state is very similar to that in the nucleus of an atom. So progress in our understanding of these 'neutron stars' was contingent upon our being sure of what laws of physics apply to them. In the case of the big-bang universe, we are completely unsure of what physics to use under the early extreme conditions.

Hoyle was bothered by a specific physical problem: where and how did the matter present in the universe come into existence? The big-bang assertion that it all came at the time of the primeval explosion begged this question since there was no physical description of the process. Indeed Hoyle felt that this was the most fundamental issue in cosmology and could not be relegated to a 'singular event'.

As we have just pointed out two papers were written on the steady-state universe. Bondi and Gold wrote theirs and sent it for publication to the *Monthly Notices of the Royal Astronomical Society* (MNRAS), under the title 'The Steady-State Theory of the Expanding Universe'. The paper was published without any difficulty. Fred Hoyle sent his paper entitled 'A New Model for the Expanding Universe' to the American physics journal *The Physical Review*, since he wished to emphasize the physical nature of his approach. The paper was rejected, for reasons we will discuss later in this chapter. He then sent it to the Royal Astronomical Society, where it was published, although it appeared later than the Bondi–Gold paper but in the same year.

Let us first approach the theory from the viewpoint of Bondi and Gold.

The perfect cosmological principle

Bondi and Gold had objected to the big-bang cosmology on the grounds that in a universe that changes so dramatically from a big bang to the present stage, we cannot be confident that the laws of physics have remained the same despite such vicissitudes. To ensure that they do not encounter any such problem in the new cosmology, they proposed to introduce a new principle which they called the *Perfect Cosmological Principle* (PCP).

We have already (in Chapter 7) come across the cosmological principle (CP) which forms the basis of most cosmological models including all of the big-bang kind. In what way was the principle proposed by Bondi and Gold 'perfect'?

Bondi and Gold wanted to ensure that we can be confident that the laws of physics remain the same at all times. To ensure this circumstance they argued that since the laws of physics are part of the universe (which, by definition contains everything in existence!), their permanence will be guaranteed if the universe remains unchanged. The PCP ensures this by stating that the universe is the same at *all cosmic epochs*. Recall that the CP ensured that the universe is the same at any given cosmic epoch; the PCP goes further by ensuring that the universe not only looks the same spatially at any time epoch, but also that it looks the same at all epochs. It is 'perfect' because it extends the concept of symmetry across space (at a given time) to symmetry across time also.

The name 'steady state' given to such a universe is then justified, because the universe looks the same at all times. If we see it today in a certain physical state, then we are assured by the PCP that it was in the same state ten or a hundred billion years ago. It would thus have the same average density and same temperature in the past as it has today. Contrast this with the big-bang universe, which would tell us that in the past the universe was denser and hotter than it is today.

The PCP is a powerful statement since it tells us that if the universe is in steady state, then you effectively know what it was like in the past and what it will be like in the future, simply by looking at its current state now. We will show how this helps us in deducing the properties of the universe.

First of all we note that such a universe cannot have a beginning. As required by the PCP, the universe exists at present and so it must have existed at all epochs in the past and it must exist at all epochs in the future. Thus, it has to be a universe without a beginning and without an end.

Next consider the Olbers paradox which we discussed in Chapter 6. The sky is dark at night. This is a fact which needs to be explained. Does it tell us about the large-scale behaviour of the universe? For example, if the PCP holds, then we have three possibilities: (1) the universe is static, (2) the universe is expanding and (3) the universe is contracting. As we saw in Chapter 6, a universe which is static and exists for ever with stars and galaxies shining within it, will be too hot and bright and so the first possibility is ruled out. The third possibility is also ruled out...because in a contracting universe, all distant galaxies are approaching us and so their radiation will be *blueshifted*. That is, the frequency of light emitted by a galaxy will increase in frequency and hence in energy. This will create a situation in our neighbourhood which will be worse than that in a static universe...the sky will be infinitely bright. The only possibility thus left is (2) which does give us a night sky that is quite dark. The radiation from distant stars is highly redshifted and so makes negligible contribution to the sky background.

So the PCP enables us to deduce that the universe must be expanding. What Bondi and Gold delighted in emphasizing was the fact that such a deduction could be made *without any telescopic observations of galaxies*. In short, it is a prediction that an armchair theoretician can make, for an observer to verify.

However, if we follow this principle further, we also deduce that in such a universe *new matter must be continually created*. For, imagine an expanding universe with galaxies all receding from us and we count the number of galaxies in a fixed volume, say, in a sphere with radius a billion light years, now and a billion years later. When we compare the two observations, we expect that the number of galaxies in the future will be smaller...as many galaxies formerly in the sphere will have moved out of it. But such a conclusion is contrary to the PCP which decrees that the number of galaxies in a given

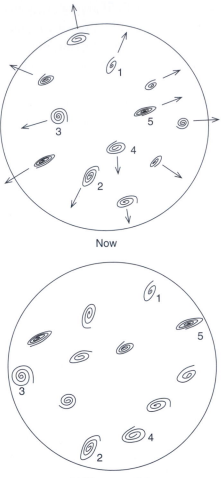

Now

1 billion years later

Fig. 9.3. The state of the distribution of galaxies at two different epochs as seen by a typical observer at the centre in the steady-state universe. The galaxies in the first figure will have moved further away from one another (and the observer) in the second figure. Some will have moved outside the horizon of the observer (shown by the round boundary). To replace those that are thus 'lost from view', new galaxies are born within the horizon sphere. Thus, on the large scale, the second picture looks the same as the first one.

volume should stay the same. So how do we resolve this conundrum? The only way to resolve it is to suppose that as the old galaxies move away new ones appear in their place so that the number of galaxies measured in the sphere stays constant.

In general we know from observations that if a gas expands, its density falls. If you take a rubber sheet and pull it outwards from all directions, it becomes thinner. These examples are generally illustrative of the maxim that matter is conserved in nature...that we cannot destroy or create it. Chemists in the nineteenth century had arrived at this law after studies of many chemical reactions that transformed matter from one form to another. Despite such transformations, they found that the total mass of all matter participating in the reaction was the same after the reaction, as before it. After the advent of Einstein's special theory of relativity, physicists learnt that this law exists in a more general form, allowing matter to be replaced by energy and vice versa. The Einstein equation $E = Mc^2$ tells us how to relate the 'lost' mass to 'created' energy. For example, in a hydrogen bomb, four nuclei of hydrogen atoms fuse together to form one nucleus of a helium atom in a thermonuclear reaction. The mass of helium made is *less* than the mass of hydrogen that was used in fusion by approximately 0.7%. What happens to the lost mass M? It appears as energy, $E = Mc^2$ as given by Einstein's formula. The quantity c is the speed of light. It has a large value and it tells us that huge energy is created by the expenditure of a relatively small amount of matter. For example, the amount of energy created by destroying one gram of matter is as much as ninety million million Joules.

So the PCP tells us that in the steady-state universe new matter is being created continuously. How is this new matter appearing in space? If we go by Einstein's formula mentioned above, we will be tempted to argue that this matter is appearing at the expense of energy. However, at first sight this does not help! There are two different effects that go against such a process. Firstly, if this universe had such a reservoir of energy it will soon be exhausted creating new matter. Secondly, like matter, energy will also be diluted by expansion and its density will fall.

A simple book-keeping exercise will illustrate this problem. Suppose we start with 10 units of energy in a given volume today. At a future epoch (say, a billion years later!) the space has expanded and so the amount of matter in the original volume is diluted down to 9 units of energy. In the meantime, one unit of energy was used to create new matter. So we will be left with only 8 units of energy. The PCP requires that the units of energy left at the end of a billion years be the same as now, that is, 10 units. So if we adhere to the law of conservation of matter and energy, then the PCP apparently cannot be sustained, and vice versa. If the PCP holds, then the law of conservation of matter and energy will have to go.

That was the conclusion that Bondi and Gold came to. They pointed out that the violation of the law of conservation of matter and energy was extremely small, and was not detectable in the laboratory. To estimate it, divide a mass of one kilogram into ten million billion billion billion billion parts. That amount of new matter per part should appear out of nothing in a volume of a cubic metre in a second.

Small though this violation was, sceptics pointed their finger at it to argue that the new cosmology was defective. Bondi and Gold argued that in the big-bang cosmology, the violation of the law of conservation of matter and energy was even more extreme. At the big-bang epoch, the entire universe appeared out of nothing. And it appears in a state of infinite density and in a manner that defies physical investigation. The apparent violation of the conservation law in the steady-state theory was, by contrast, very mild, and could be studied by astronomical observations.

Bondi and Gold did not go beyond the PCP to lay down quantitatively a theoretical framework analogous to that of the big-bang cosmology whose mathematical models follow from Einstein's equations of general relativity. They felt that the PCP provided ample ways of testing the predictions of the theory. For example, if we observationally measure the density of matter in the universe today in our neighbourhood, then we can assert that the same density existed in the past epochs, and this prediction can be tested.

As Hermann Bondi was at pains to emphasize, the steady-state theory satisfied Karl Popper's criterion of how a physical theory should be. The theory should make predictions that can in principle be tested and disproved. The steady-state theory made definitive predictions and this made it vulnerable to observational tests. This was in a sense also the great strength of this cosmology. As we shall see, because of its reputation as a theory that makes clear-cut predictions, the steady-state theory had to bear the brunt of several attacks, some fair, some unfair.

Creation of matter

Fred Hoyle approached the steady-state theory from a physicist's point of view rather than from the empirical philosophical point of view of Bondi and Gold. He realized that the steady-state theory would be called upon to explain how the new matter was created. Was the sacrosanct law of conservation of matter and energy violated in this cosmology? Even if one pointed the finger at big-bang cosmology as the greater offender, like Caesar's wife, a scientific theory needs to be blameless. Hoyle's approach was therefore to go back to Einstein's equations (which had served as the launching pad for the big-bang cosmology) and see if they could be modified to permit a 'non-big-bang' type solution. That is, a model of the universe without a singular epoch in which the mystery of creation of matter can be understood.

In his original paper of 1948 Hoyle found a solution to this problem, although in formal terms it was improved upon by his friend and mentor Maurice Pryce who never published his work. We try to capture here the basic idea behind these approaches.

The trick lies in introducing a new field of *negative energy*. The 'field' is a concept popular amongst physicists for describing entities that carry physical effects from one place to another and which require energy to do so. The basic idea came with the notion of the electric and magnetic field popularized by Michael Faraday in the nineteenth century.

A simple experiment will illustrate the idea. If we place a bar magnet on a white piece of paper, all that we are aware of is the magnet and the paper. However, unseen to the eye there is also present the magnetic force of the bar magnet. We can experience this force by bringing a small ball of iron near to it. The ball will be attracted by the magnet. If we do the experiment carefully, we will notice that the ball moves in a certain curve towards the magnetic pole. An indication of this will be seen by sprinkling iron filings on the paper. Then the filings align themselves in a certain pattern (shown in the figure below). The lines along which the filings lie are often called the lines of force and the directions of these lines are the directions of magnetic attraction. These lines are indicative of the fact that the region near the magnet is not empty (as it may

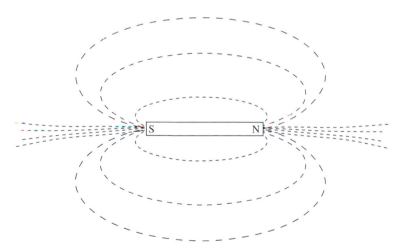

Fig. 9.4. Following Faraday, we describe here the concept of magnetic lines of force. If we spread tiny iron filings near a bar magnet then they become aligned along curved lines starting from one pole of the magnet and ending at the other, as shown in the figure. The typical line denotes the direction of the magnetic force acting at its points. The concept of lines of force is formalized in the notion of the 'magnetic field'. It is an entity filling space and its strength and direction at any point is measured by the magnetic force at that point.

appear to the eye), but contains a record of the magnetic forces originating in the magnet. The intensity is higher close to the magnet and lower away from it. The fact that it can move a ball or iron filings indicates that the field has energy.

Normally physical fields have positive energy. When such a system interacts with matter and moves it (as in the case of the iron ball moved by the magnetic field) it has to expend energy, and so its energy reservoir *gets depleted*. The process works for as long as the field has energy left within it.

What would happen if fields with negative energy were also present? In such a case expenditure of energy used to move matter will make a field *more negative*. Imagine, we have an energy reservoir containing -10 (minus ten) units of energy. One unit was spent in moving matter. What is left? The balance is now -11, that is a quantity whose magnitude is greater than before. An analogy to everyday life will help us to understand this situation.

Consider two individuals A and B. A has a savings account in his bank and it contains ten units of currency, while B has an overdraft on his bank account amounting to 10 units of currency. That means B owes the bank 10 units. Whenever A spends money from his account his balance decreases and when it reaches zero, he will not be allowed by his bank to spend any more on that account. What about B? Suppose he decides to go on a shopping spree and spends one further unit of currency from his account. If the bank allows him to do this, B's overdraft will have increased to 11 units of currency. If the bank does not put a limit on overdraft, B will go on spending recklessly. By placing a ceiling on B's overdraft his bank does not allow such reckless spending.

Just as reckless spending is not allowed by the bank, so physicists normally assume that nature does not allow negative energy fields to exist because this would lead to uncontrolled movements and expenditure of energy. Indeed this restriction seems to be obeyed by almost all forces in nature. 'Almost' because, we know of one striking exception. The force of gravity has this strange property. The more one draws upon it to move matter the stronger it grows.

Fred Hoyle's idea of using a negative energy field for describing creation of matter can be appreciated by recalling our earlier example of why a positive energy field does not work. A field of positive energy becomes steadily depleted in strength because of two effects: (i) the expenditure of energy used to create matter, and (ii) the expansion of space as the universe gets older. Now take a negative energy field whose reservoir has -10 units of energy in a given volume. Because of the first effect, the creation of matter, the energy is reduced to -11 units, say. However, the second effect of dilution by expansion reduces the magnitude of this reservoir, changing it back to -10 units. So we are back to where we started, but at the same time we managed to produce one unit of

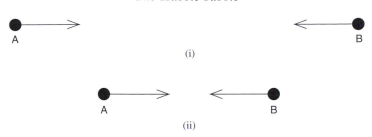

Fig. 9.5. The two masses shown above, are attracted by their force of gravity. If we allow them to come closer, as required by that force, what happens? The force of gravity grows. And this growth is, in principle, unlimited. By way of contrast, imagine two masses held by a spring that has been stretched. These masses feel the force of the spring pulling them closer. What happens when the masses come closer? The force of the spring disappears, as it is no longer stretched.

new matter! This may appear to be sleight of hand, but is really quite a sound idea mathematically. This is how, by adding a negative energy field, called the *C*-field, to the original Einstein equations, Hoyle was able to get the very same steady-state model that Bondi and Gold had proposed from their perfect cosmological principle. He was also able to show that in the process of the creation of new matter, the law of conservation of matter and energy is not violated.

Nevertheless, in 1948 the physics community was not prepared to accept negative energy fields. This is why Hoyle's paper did not find favour with the physics journal. Ironically, today, 60 years later, negative energy fields are getting popular with the physics community and even in cosmology. We shall return to this situation in Chapter 14.

Since Hoyle developed a quantitative theory he was able to relate the dynamical properties of the universe to the basic constants of gravitation and the *C*-field. For example, the density of matter in the universe turns out to be *twice* the critical or closure density of the big-bang cosmology *at the present epoch*. Because the steady-state model has the same rate of expansion (Hubble's constant) at all epochs its density also remains constant at all times. It is also possible to demonstrate, as was done by Hoyle and Jayant Narlikar in 1962, that the steady-state model is *dynamically stable*. That is, it returns to its standard form even if disturbed by some cosmic event at any time.

The Hubble bubble

Hoyle's approach improved upon by Maurice Pryce led to a few studies of the universe in the early 1960s which turned out to be prophetic. We briefly describe these ideas by Hoyle and Narlikar using the *C*-field. All three concepts came before their time, they were opposed at the time, but later they became part of mainstream thinking.

First, the notion that baryons need not be conserved was accepted by particle physicists. By baryons, we mean the heavy particles like neutrons and protons that make the atomic nucleus. The steady-state theory suggested that the newly created matter was mainly in the form of baryons which later made up larger conglomerations and formed galaxies. Particle physicists were firmly opposed to such a suggestion, arguing that the 'baryon number is a conserved quantity', meaning that the universe came with a fixed baryon number and no interaction creating or destroying baryons could alter it.

However, by the end of the 1970s, the classical big-bang cosmology also came around to this view! First it was found that if one expected complete symmetry between matter and antimatter in the universe, then we have the troublesome problem of understanding where all that antimatter is. Rather than explain it away, cosmologists felt that perhaps basic interactions in

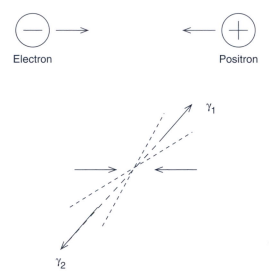

Fig. 9.6. Antimatter is in a sense the symmetric opposite of matter, whose existence was required by the quantum theory of matter, and was indeed confirmed experimentally in the 1930s. For example, the subatomic particle, the electron, has an antimatter counterpart called the *positron*, with the same mass but with equal and opposite electric charge. When, as seen in this figure, an electron meets a positron, the two annihilate each other producing energy in the form of radiation. This is typical of all matter–antimatter pairs. Physicists believe that nature has matter and antimatter in equal proportions. But then our observations show the universe to be largely made of matter. What happened to the antimatter? If the universe was created in a big bang, then when and how was the symmetry between matter and antimatter broken in favour of the former? This is the question that cosmologists must answer.

physics very early in the universe would suggest a way out. Their collaboration with particle physicists described in Chapter 13 led to the same conclusion. For, the grand unified theories speculated on by particle physicists led to the conclusion that at some stage the matter–antimatter symmetry was lost and more baryons than antibaryons were created. In short, from being an anathema of the 1960s, the violation of baryon number became acceptable in the 1980s.

The second aspect of the Hoyle–Narlikar theory was that there was possible evolution of the steady-state universe in which matter creation may be suppressed locally. In such a case that region expands as a Friedmann–Lemaître

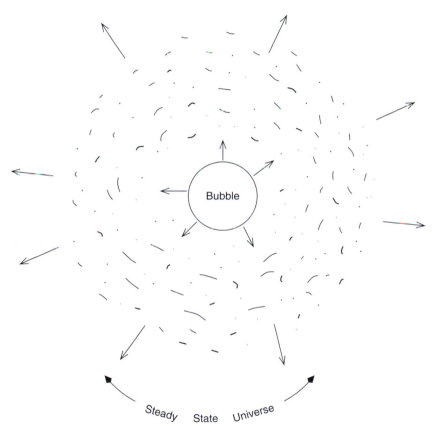

Fig. 9.7. In the Hoyle–Narlikar cosmology, the idea was proposed that the universe is always in a steady state with high density everywhere, but in some parts the creation activity was switched off and the region expanded as a Friedmann model. This would lead to the region appearing as a low-density bubble in a denser medium. Sometimes this was called the 'Hubble bubble', since Hoyle and Narlikar had argued that the normal expanding universe picture arising from Hubble's observations related to this bubble-like region.

universe. Such a universe will be seen as a tiny bubble within a vastly bigger steady-state universe outside. The bubble was sometimes referred to as the 'Hubble Bubble'. In 1981, Alan Guth proposed an inflationary universe along similar lines. But in his case the basic theory was not specified but believed to be a grand unified theory of the kind talked about by particle physicists (*see* Chapter 13). Ironically, 'inflation', a concept currently very popular with big-bang cosmologists was anticipated twenty years earlier by the steady-state cosmology.

The third prediction of the steady-state model was that there may be locations where matter is created in large lumps rather than in a homogeneous way. A typical large lump was believed to exist as a massive object whose gravity would be so strong that even light might have difficulty escaping from its surface. Later such objects became known as 'black holes'. The idea was proposed by Hoyle and Narlikar, that these objects would serve as nuclei of galaxies and argued that typically elliptical galaxies should have such massive objects in their nuclear regions. In the 1960s, the current thinking ran against this suggestion, but today it has become well established.

Response of the academic community

Hermann Bondi once said that a theory is 100% successful if everybody accepts it, but it is 50% successful if it is opposed and debated, and its success record is 0% if it does not excite any comment, positive or negative. The steady-state model in both of its versions can be said to have scored 50% on Bondi's success scale. The philosophical approach of Bondi and Gold drew ire from some philosophers. An outburst by Herbert Dingle almost verged on the hysterical. In his presidential address to the Royal Astronomical Society, Dingle made the following harsh pronouncement on the steady-state theory, especially the PCP of Bondi and Gold:

I do not enjoy the task of arraigning those whose mathematical facility greatly exceeds their judgement of scientific authenticity, and who have in consequence exercised this facility on any premises that will give it scope. But one who, however unworthy, accepts the honor of presiding over one of the foremost scientific societies of the world, accepts a responsibility. The ideas to which we give publicity are accepted as genuine scientific pronouncements and as such influence the thinking of philosophers, theologians, and all who realize that in no intellectual problem, however fundamental, can scientific research now be ignored. And so when it happens we have published, in the name of science, so-called 'principles' that in origin and character are identical with the 'principles' that all celestial movements are circular and all celestial bodies immutable, it becomes my duty to point out that this is precisely the kind of celebration that science was created to displace.

Physicists were generally hostile because of the apparent violation of conservation of matter and energy (in the Bondi–Gold model) or the use of a negative energy field (in Hoyle's version). However, many intellectuals were attracted to the steady-state idea since they were unhappy with the singular origin required in the big-bang cosmology. By and large astronomers kept an open mind and liked to take up Bondi's challenge of disproving the PCP or the steady-state theory by direct astronomical tests. We shall describe some important contributions made to observational astronomy by such attempts.

10

The cosmological debate 1950–1965

Attacks on the steady-state theory

Despite their claim of objectivity, most scientists are influenced subtly in their thinking by their cultural background. Since, in the twentieth century the developments we have been describing mostly took place in the Judaeo-Christian world, these religions had their effects on cosmological thinking.

There are not many of us, apparently, who can keep their faith apart from their science. However, Georges Lemaître was one who could. He had been the originator of the big-bang model, at least in drawing attention to its singular beginning, by labelling it as the 'primeval atom'. Once, in the late 1950s, while attending the Solvay Conference in Belgium (an international gathering of scientists whose names could very well appear in 'Who's Who'), he was asked by the radio astronomer Bernard Lovell 'Georges, how do you think it all started?' To which Lemaître replied: 'As a priest I can tell you; as a scientist I can't.'

Unfortunately, this openness of thinking is missing in most cosmologists, many of whom hailed the 'blessing' given by the Pope to the big-bang scenario at the Rome meeting of the International Astronomical Union in 1952. The idea of a 'divine act of creation' in the Old and New Testaments comes close to the big-bang concept and, to many scientists, justifies the singularity that goes with it. To them, the idea of a universe without a beginning and without an end does not make sense and hence there was and has been a cultural antipathy to the steady-state concept amongst Western scientists and philosophers.

However, in the arena of science, it is not the cultural prejudice that should hold sway, but rather what the facts are. What real evidence do we have that the universe indeed had a past history different from the present as is suggested by the big-bang model? In this book we shall be concerned with assessment of facts, such as they are, to answer this question.

Because the steady-state universe made definitive predictions, observers attempted to disprove the theory by making direct measurements of the universe to show those predictions to be wrong. As mentioned above, the attitude was not the objective one, to ask: 'What do the facts tell us about the past history of the universe?' Rather it boiled down to 'The steady-state theory is bound to be wrong: let us find ways of demonstrating this result.' As a result, there were many claims to have disproved the model, claims that under closer examination came to nothing. We will discuss some important episodes of this nature in this chapter.

The Stebbins–Whitford effect

Within a few years of the papers by Bondi, Gold and Hoyle proposing the steady-state theory of the universe, a claim was made that the observation of Joel Stebbins and Albert Whitford showed that the energy distribution of relatively distant galaxies was much redder than the same energy distribution in nearby galaxies. This issue of 'far' and 'near' will feature frequently in our discussion here and so we first state its implications.

Whenever we observe a distant galaxy, we see it as it was in the past, not what it is like today. Suppose we see a galaxy a hundred million light years away. What does this distance mean? The distance of that galaxy is so large

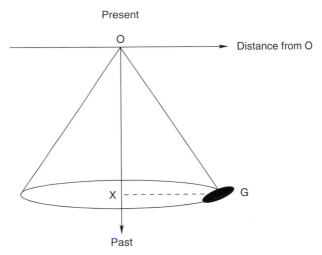

Fig. 10.1. This cone-like diagram shows that as we observe more remote regions of the universe we look back in time. This is because light takes time to travel and this 'past light cone' shows how light rays left remote regions at past epochs so as to arrive at the observer at the present epoch. So a distant galaxy G may appear young because it was young at the time this light left it; but at the present epoch it will be quite old!

that it takes light a hundred million years to travel to us. This means that the light which carries information to form the image of the galaxy on our photographic plate (or, the CCD detector) left the galaxy a hundred million years ago. In short the information on which our image is based is a hundred million years old. So we are seeing the galaxy as it *was* rather than as it is now. Imagine an extraterrestrial individual on a planet going around a star about 340 light years away looking at the Earth with a powerful telescope. He is able to observe the Earth as it was 340 years ago...seeing Isaac Newton working in England and Emperor Shah Jahan building the Taj Mahal in India.

Like this imaginary glimpse into history, the astronomer is rewarded with glimpses of the past when he looks at galaxies further and further away. A photographic plate contains images of stars and galaxies from far and near. Unless the observer is cautious, he may make mistakes of interpretation. The simplest one would be explicable if we use the example of a composite family photograph which has members of several generations of a family together. Now, suppose in this day and age of image processing we prepare a composite photograph in which the different members have been photographed at different times. We may have the grandfather appear as a teenager because his photograph was taken fifty years ago. His son, the father photographed today, will therefore look older than the grandfather.

This situation of course, does not happen in everyday group photographs but can often occur naturally for an astronomer. Thus he may see a distant galaxy, which may look very young at the time the light left it heading for our camera. By contrast, a nearby star may look much older, whereas the true situation would be the reverse: *today that galaxy will be older than the star*. In short we must recognize the distance factor while dating objects as they are today.

In cosmology, distances are generally estimated on the assumption that Hubble's law holds for all galaxies. In Chapter 16 we will present evidence that casts doubt on this widely held hypothesis for some classes of objects. However, for the present, we will assume that the larger the redshift of the object, the further away it is and, the further back in the past it is observed. Therefore, if the universe is in steady-state, then for any population of objects, the overall properties should be the same at all epochs, i.e., they should not depend on the redshifts at all.

This simple conclusion drawn from the *Perfect Cosmological Principle* (PCP; see previous chapter) illustrates the predictive power of the steady-state theory. One example of this conclusion relates to the age distribution of galaxies. Today we expect to find galaxies of all different ages in our neighbourhood. The older ones will be rarer because of the expansion of the universe: they will have moved away from us over the period they have been in existence. The younger galaxies will be seen in greater abundance since they were created recently and have not had time to move far away. But more importantly consider a sample of galaxies located far away with larger redshift, which belongs to a past epoch. The properties of these galaxies should be the same as those found in our neighbourhood. In particular, the age-distribution of galaxies should be the same.

To test this hypothesis, Stebbins and Whitford looked at the colours of galaxies, some in our neighbourhood and some, at larger redshifts. They found that as the redshift increased, the colour showed an increasing redness. In short there was a progressive reddening of galaxies with increasing redshift. On the face of it, this was evidence that there is a progressive change in a physical property (namely, colour) of the galaxies as we probe them further into the past. If we relate them to the ageing of the stars in the galaxies, this could be explained by saying that the higher redshift galaxies are progressively younger.

Since this went directly against the PCP, it was often cited as evidence against the PCP. However, theoretical analysis of the effect, showed that a subtle issue had been missed out. We try to explain it next.

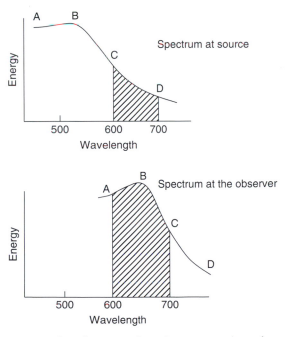

Fig. 10.2. The spectrum of a distant galaxy is compared at the galaxy and at the observer. Because of the redshift, the spectral features move towards longer wavelength as shown in the two spectral distributions.

Suppose an astronomer is looking at a spectrum of a nearby galaxy. The spectrum shows how much energy is being radiated by the galaxy in any given range of wavelengths. The red colour corresponds to the wavelength range 600 to 800 nanometres. We suppose also that the energy per nanometre width increases as the wavelength decreases below 600 nanometres. So, a range of 500 to 600 nanometres (corresponding to other colours like green, yellow, orange) has greater energy than the red colour over the band 600–700 nanometres.

Now suppose that we are looking at a very similar galaxy at a redshift of 0.2. Because of the effect of the redshift, the wavelengths received by the astronomer on Earth will be longer than the wavelengths emitted by the galaxy. So what is being seen as a 'red' colour here may have been yellow or orange or even green at the source. And because the galaxy emits more energy in these colours, they will show up as red with much greater energy. Thus, even though the galaxy population has not changed in its spectral properties, it may show up as progressively redder at larger redshifts.

Only in 1968, did Allan Sandage and Beverley Oke quantitatively demonstrate this to be the real cause of the Stebbins–Whitford effect. They showed that after this redshift effect on the galaxy spectrum was taken out, there was nothing significant left to be explained by a stellar ageing process. In short, the PCP was consistent with what was observed, at least up to the redshift of 0.2, which was the limit of the Sandage–Oke sample.

While this explanation of the Stebbins–Whitford effect was known in the 1950s and Bondi had forcefully argued that the effect had failed in disproving the steady-state theory, quantitative observational proof had, however, to wait till 1968. Meanwhile many astronomers still had the mistaken impression that the Stebbins–Whitford effect showed that the steady-state theory was wrong.

The counting of radio sources

We have noted in an earlier chapter (Chapter 8) that Hubble's programme of counting galaxies out to different distances in order to decide the geometry of the universe was eventually abandoned, but was taken up in a modified form later by radio astronomers. By the mid 1950s the controversial issue as to whether the majority of strong radio sources were galactic or extragalactic had been resolved. They were extragalactic. Now radio astronomers believed that the radio sources they were observing went far beyond the galaxies that the optical astronomers could easily detect on their photographic plates. Also the number of radio sources to be counted was much less than the number of galaxies. So, prima facie, it made sense to use the radio sources as the objects to be counted out to different distances.

However, there were problems in doing this which were peculiar to radio astronomy itself. The most crucial was the inability of radio astronomers to appeal to Hubble's law in order to estimate the distances of radio sources. If the redshift is z, then by Hubble's law, the distance is as $c \times z \div H$, where H is the Hubble constant. The radio techniques did not provide the redshift of a galaxy.

Using the detector at the focus of the radio telescope we can measure how much radiation is coming from it in a given time (usually one second) over a given area (usually one square metre) in a given bandwidth of frequencies (usually one Hertz). The further away the radio source is the smaller is this quantity, usually called the 'flux density'. If the amount of energy is one watt, then the flux density it corresponds to is one Jansky. (Calling the unit a *Jansky* commemorates the pioneering achievements of Karl Jansky in radio astronomy, described in Chapter 8.) Thus the flux density is often taken as indicative of the distance of radio sources.

For stars also a similar distance indicator is used (*see* Chapter 3). The inverse square law of illumination is used for comparing distances of two similar stars. Following the same argument we can say that if we have two radio sources A and B with the flux density of A as 100 Jansky and the flux density of B as only 1 Jansky, then this law suggests that B is 10 times further away than A. For, $100 = 10 \times 10$: the ratio of intensities (100) is the square of the ratio of the distances (10).

This rule only works *if we are sure* that the radio sources A and B are similar in the amount of radio energy that they emit per second. Experience shows that radio sources vary a lot in their physical characteristics including the energy that they emit. So, there is the danger of our misinterpreting the data. If in the above example, B is intrinsically 100 times weaker compared with A, then in reality both A and B will lie at about the same distance from us.

The stellar astronomer takes care to avoid a similar mistake by ensuring that the stars he is observing are very similar in their physical characteristics. Because considerable progress has been made in understanding the physical characteristics of stars, such mistakes are rare. But without a proper understanding of the physics of the radio source, it is therefore dangerous to use this method to estimate distances.

A better way is to optically identify a radio source; that is, to find the optical image of the galaxy which is giving rise to the radio source. This was done by Walter Baade for Cygnus A (*see* Chapter 8). Once we have identified the galaxy we can obtain its spectrum and determine its redshift and then use Hubble's law to estimate its distance. While this technique is better, it won't work unless the optical source can be identified. In other words, the radio astronomer can be sure of the distance estimates only for those sources that the optical astronomer can also reach.

These issues were glossed over in the early enthusiasm of radio astronomers for their new techniques. Apart from Hubble's original programme of determining the geometry of the universe, there was the steady-state model to shoot at. Can the counting of radio sources demonstrate that there were more radio sources per unit volume in the past than there are now? If so, the PCP would be disproved and along with it the steady-state model.

Martin Ryle who led radio astronomy at Cambridge, was particularly keen to have a go at the steady-state theory. His reasons may not have been entirely objective. We have already mentioned his debate with Tommy Gold, which he had lost (Chapter 8). It would have been sweet revenge if by counting radio sources he could disprove Gold's steady-state theory. Apart from Gold, the two other authors of the theory were not exactly friendly with Ryle. In particular there existed considerable animosity between Fred Hoyle and Martin Ryle over many issues, both astronomical and otherwise. The second motivator was Ryle's cultural Christian upbringing. The big-bang concept naturally appealed to him in preference to the steady-state theory.

Bernie Mills, a pioneer of radio astronomy in Australia, also undertook a similar survey, although he was not prejudiced in favour of the big-bang theory. He simply wanted to count the sources and to look for any deviation from Euclidean geometry, as Hubble had tried to do for galaxies. By counting radio sources to different and progressively decreasing levels of flux density one can check this hypothesis. Mathematical analysis shows, for example, that if the number of radio sources with flux density greater than 1000 Jansky is 5, then the number of such sources with flux density greater than 10 Jansky should be 5000. If the Euclidean hypothesis is wrong, then the number at 10 Jansky should be significantly greater or less than 5000.

The survey of Mills and his colleagues indicated numbers somewhat larger than the Euclidean hypothesis allows. However, Mills and his colleagues concluded that the difference was not significant. That is, their conclusion was a cautious one that the counting carried out by them did not reach the level where a cosmological judgement could be made.

On the other hand Ryle's conclusions were very different. His 1955 survey indicated a sharp rise above the Euclidean values. In our example above, the number corresponding to 5000 would have been as high as 5 000 000. He claimed that this steep rise was not only in disagreement with the Euclidean universe, but also disproved the steady-state theory. Two years later Ryle and his group announced the results of another larger survey which also disagreed with the steady-state theory and also with a Euclidean universe. However, now the discrepancy with the theory was much less than before. In our example above, instead of the number at 10 Jansky being 5 000 000, it came down to 50 000.

Why was there such a difference from the previous survey? It turned out that the earlier survey had several errors in measurements not appreciated at the time. It was claimed that these were now under control and the new result was sound.

When one is observing at the limits of the instrument, there are bound to be errors in measurements. These need to be understood and estimated in order to attach credibility to the conclusions drawn from the data. This had not been done adequately in the first survey, nor in the second one. At this time the Australians announced their result which we have mentioned before. This result differed from the Cambridge result and the reasons needed to be understood.

Because of their different geographical locations, the Cambridge group had been observing radio sources in the northern part of the sky, while the Australians were looking at the southern part. Still, there was some small overlap and therefore a comparison of the two surveys could be made. It turned out that the Cambridge measurements were in error, but Ryle and his group did not accept this interpretation, and maintained that the Australians were wrong. Ryle, meanwhile, had embarked on an even greater survey which went considerably deeper into space by looking at much fainter sources. The results of this survey were expected towards the end of 1960.

The Ryle–Hoyle controversy

It was in the summer of 1960 that one of us (Jayant Narlikar) embarked on his research career as a Ph.D. student of Fred Hoyle. When he went to see Hoyle for his first briefing, Hoyle suggested a number of possible research problems in theoretical astrophysics and cosmology. However, the steady-state theory did not figure in the list. Why? As Fred explained: 'This is a controversial theory and I do not think a new research student should work on a controversial topic.' Narlikar went back rather dissatisfied that he was denied working on a theory he liked for its simplicity.

However, six months later the situation had changed dramatically. In early January, 1961, the Mullard Company (which had helped build Ryle's radio astronomy observatory) invited Hoyle to a press conference, at which Ryle was to announce his new results. These were from what was known as the 4C

Survey. (The Cambridge surveys were numbered 1C, 2C, 3C, etc. We have just been describing the 2C and 3C surveys.) The announcement turned out to be a denunciation of the steady-state theory based on the results of the 4C Survey. Hoyle, as one of the originators of the theory, was immediately asked to respond. He found himself at a strategic disadvantage since he had not been given the data on which the claim of disproof of the steady-state theory was based. He realized belatedly that he had been set up. The press conference resulted in considerable publicity for Ryle and Mullard, with the headline that the big-bang theory was established once and for all.

Hoyle then felt that a suitable reply to Ryle must be given, not through the media, but at the Royal Astronomical Society where Ryle was expected to present his technical report. Time was short, since the RAS meeting was scheduled for the second Friday in February, that is, on February 10. In the three weeks or so remaining he needed to study the data and come with a scenario that would demonstrate that even if the data were correct, the steady-state theory was not disproved by it.

This is when Hoyle and Narlikar had to work out the details of a model from the skeletal framework that Hoyle had. Narlikar was at last given the opportunity to work on the steady-state theory. He recalls working together with primitive computer facilities to explore the theoretical framework. But the greater difficulty lay in getting the data out of Ryle's group.

Ryle was well known to be very secretive. He always discussed his ideas and results in closed meetings involving the people he worked with and if any of them discussed results outside the group they would be expelled. He behaved this way to exclude other radio astronomy groups, theorists, and above all, anyone from the US.

In this case, Hoyle could not reply to Ryle's arguments unless he was given access to the facts, the data. After great difficulty, Ryle agreed to see Hoyle and Narlikar in the tea room at the Cavendish Laboratory. He had brought with him a sheet of paper on which the number count–flux density curve was plotted by hand, with some numbers scribbled on it. Fit that curve to your theory if you can, was his challenge. No details were available regarding observational errors, telescope limitations, etc., let alone any catalogue giving details about the sources counted. Such data were never provided to outsiders, particularly those who might be critical.

Here we can discuss these results using the example of the number counts given earlier. Recall that if the number of sources brighter than 1000 Jansky was 5, in the Euclidean theory we would expect there to be 5000 sources brighter than 10 Jansky. What Ryle was finding was a number like 20 000. Although this was a further decrease below earlier estimates of 5 000 000 and

50 000, it was still four times the expected value in a Euclidean universe. (For a comparison, the Australians were finding a number around 10 000, which they discounted as not significantly different from the expected number in view of the likely errors of measurement.) Furthermore, Ryle claimed that the rise continued to lower flux density, as low as 2 Jansky.

To explain these data, Hoyle introduced a new element into the picture, namely that of inhomogeneities on a larger scale than had previously been thought to be present in the universe. He assumed that there are regions as large as 150 million light years populated with galaxies, separated by voids equally large. These populated blobs contained a large number of clusters of galaxies and hence could be called *superclusters*. In fact Hoyle and Gold had proposed a theory of galaxy formation in the steady-state universe wherein galaxies form in large regions like the superclusters. Given inhomogeneities on this scale, a typical observer may see different number count curves from different locations. Indeed, a year later, when the facility of the IBM 7090 computer became available in London, Narlikar devised a Monte Carlo type programme to create randomly placed observers in the steady-state universe containing superclusters and voids on this scale. He found that, depending on the location, the observer would get counts exceeding or less than the Euclidean prediction.

By the end of January, Hoyle and Narlikar were able to demonstrate that the observers could get higher numbers than the Euclidean prediction, if the universe was inhomogeneous on this scale. In short, *it was possible to get Ryle-type number counts in the steady-state universe*. The Royal Astronomical Society had allocated ten minutes to Hoyle for presenting his response to Ryle on February 10. But Hoyle had a conflict of schedules on that day. He was already committed to address an undergraduate audience in London that afternoon, an engagement he could not get out of. So how would his reply to Ryle be presented? It would have to be given by his student and co-worker, Narlikar.

The original intention of Hoyle had been to keep his student protected from controversy by denying him the opportunity to work on the steady-state model. Now he was right in the midst of the controversy, being asked to defend the model against a seasoned critic like Martin Ryle. Whatever inhibitions he had in taking part in the debate were set at rest by the assurance of his mentor who trained him in the art of presenting the crux of the argument in eight minutes...leaving two minutes for unforeseen interruptions. In the end Narlikar did well, judging by the feedback he received from individual members of the packed audience.

Thus, although the defence of the steady-state theory was well done, it was achieved at a cost, that of supposing that the universe is much more

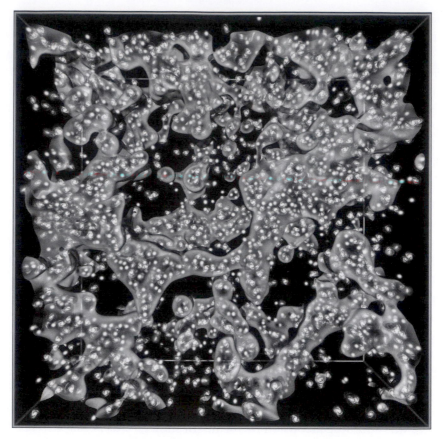

Fig. 10.3. The typical structure in the three-dimensional distribution of galaxies in the universe using a 100 Mpc sized cube carved out from the 2dF Galaxy Redshift Survey (2dFGRS). The grey shading is a visual aid that highlights the density contrast in the distribution by marking a region of approximately constant density. Credit: *Paul Bourke/Swinburne Centre for Astrophysics & Supercomputing and the 2dFGRS Team.*

inhomogeneous than was widely believed at the time. Was the price paid worth it? There were already indications of higher order clustering in the universe. Gerard de Vaucouleurs had been arguing that our own Milky Way lies in the Virgo Cluster which belongs to a larger entity which he called the *Local Supercluster*. Its size is about the same as assumed by Hoyle and Narlikar. Moreover, an analysis of the distribution of galaxies in the sky by George Abell in 1958 had shown that there was not only clustering, but also super-clustering on the larger scale of 150–200 million light years. Nevertheless a large body of astronomers in the 1960s frowned upon inhomogeneities on such a large scale. It was not till the early 1980s that the idea of superclusters and voids became well established and accepted.

What was the aftermath of the Hoyle–Ryle controversy? Because of the publicity given to Ryle's work many people believed that he had demolished the steady-state theory. However, as Hoyle and others had consistently argued 'you can't base your cosmological judgment on counting of objects that you do not understand physically'. At about this time several radio astronomers and astrophysicists turned to understanding what radio sources are really like. And although source counts have continued to be made down to fainter and fainter flux-density levels (today they go down to milli-Jansky) it is realized that they do not help in explaining what type of universe we live in. The subtle geometrical differences between different models are dwarfed by the physical diversity of the radio source populations. Towards the end of this chapter we will describe some of this diversity.

The revisions of Hubble's constant

The period after 1950 also saw important developments in optical astronomy. Bringing into operation the 200-inch Hale telescope in 1948 may not have furthered Hubble's initial aim of determining the geometry of the universe by counting galaxies, but it led to advances in other directions. The most basic was the determination of Hubble's constant itself. For, by this time it had become clear that there had been serious errors in Hubble's distance estimates back in 1929. Since the Hubble law relates radial velocities to distances through Hubble's constant, it was this constant that came under review.

We can place the seriousness of the issue in proper perspective by recalling that the Hubble constant H appears in the simple formula $V = H \times D$. Here D is the distance of a galaxy and V is its radial velocity. The velocity is estimated by multiplying the speed of light c by the redshift z of the galaxy. Although we have so far used the unit, light year to measure distance, astronomers, for historical reasons have been using the unit called *parsec*. We remind the reader that the parsec is approximately three and a quarter times a light year. As a ready reckoner, we will use one parsec as three light years.

In 1929, Hubble and Humason found that the value of H was about 550 kilometres per second per megaparsec. That means, a galaxy at a distance of a million parsecs would be moving radially away from us at the speed of around 550 kilometres per second. Milton Humason was a brilliant observational astronomer. He made many of the key observations of giant galaxies attributed to Hubble and Humason. He had had no professional training and started as a construction worker when the Mount Wilson Observatory was built in the early 1900s.

As time went on a number of mistakes in using standard objects to derive distances were clarified, and gradually the value of the Hubble constant was revised downward corresponding to an increase in the distance scale and the 'age' of the universe. By 1956 the value of H had been reduced from 550 to 180, and since 1956 it has been reduced further to a value of about 60. Even today there is a serious difference of opinion as to what it actually is. Depending on who you ask, the answer may range from 40 to 75 kilometres per second per megaparsec.

Is the universe decelerating?

Theoretical models of the big-bang cosmology tell us that the value of Hubble's constant changes with epoch and may not have been the same in the past. For models *without* the cosmological constant the value of H was higher in the past than it is today. We may express this fact by saying that the universe was expanding more rapidly in the past than today, so that it is slowing down or *decelerating*. If we allow models with a positive cosmological constant, the conclusion is the exact opposite: the universe should be *accelerating* at this time. In the 1950s, the majority of cosmologists believed that there was no need for a cosmological constant (which its originator, Albert Einstein himself had disowned in the 1930s). Thus the interest was centred largely on decelerating models.

For observers to shoot at, the theorists had defined a parameter called the *deceleration parameter* and it was denoted by the symbol q. This parameter was expected to be positive. Moreover, a very simple result emerged from the mathematics of the big-bang models. If q turned out to be exactly ½, the spatial geometry of the universe is flat, i.e., Euclidean. If q were greater than ½ then the geometry would be that of a closed universe, while if q were less than ½, the universe would be open. So by measuring this parameter observers could tell what the geometry of the universe was like.

Where did the steady-state theory stand in this respect? It described a universe that is accelerating all the time. For this, the value of the deceleration parameter q is negative and equal to -1. So it stood out apart from the bunch of big-bang models. And here lay another chance of disproving the model by simply measuring q and showing it to be positive.

Allan Sandage, a leading astronomer at the Palomar Observatory and the successor to Hubble, had been the main observer following this test. Allan has not only spent decades of effort measuring the Hubble constant (as mentioned before) but has also attempted measurements of q. Figure 10.4 illustrates the outcome of one such effort in the 1960s. There we see different curves superimposed on data points. Each point plots the apparent magnitude (that is, how bright the galaxy looks from here) on the horizontal axis and its redshift on the vertical axis. A relative comparison of the curves shows that the curve for the steady-state theory moves away in the horizontal direction from the rest of the curves describing different big-bang models with decelerating expansion. Which theoretical model seems preferred by the observed distribution of points? On the basis of such a graph and the closeness of observed points, Sandage claimed that the model with a positive q, close to 1, was preferred. This meant that the steady-state theory was ruled out.

However, by the mid 1970s, it was beginning to be clear that the uncontrollable errors in making such a test were larger than estimated in the earlier decade by Sandage and his co-workers. Jim Gunn and Beverley Oke demonstrated the existence of various artifacts in the observing techniques that enlarged the error-bars. Given these larger error-bars on the possible value of q, Gunn and Oke concluded that even negative values were not ruled out. Thus the early objection to the steady-state theory seemed to have gone. Moreover, later observations by Sandage lowered the value of q to zero, thus allowing both decelerating and accelerating models.

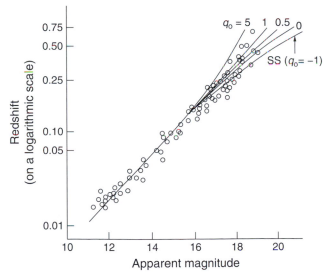

Fig. 10.4. A plot of the measurements of radial velocity (redshift) and distance (apparent magnitude) of galaxies obtained by Allan Sandage and his colleagues in the 1970s. The parameter q_0 measures the extent of deceleration of the expansion of the universe. All positive values correspond to the Friedmann models, whereas the negative value of -1 represents the steady-state model which predicts that the universe is accelerating.

Nevertheless, after seeing the evolution of this test over the years most cosmologists had begun to discount its efficacy in distinguishing between different theoretical models. The situation remained dormant for a decade and a half, when, towards the end of the 1990s, there was a surge of interest in this test. We shall return to it in Chapter 14.

The discovery of quasi-stellar objects

Shortly after their confrontation with Martin Ryle in 1961, Hoyle and Narlikar paid a visit to Robert Hanbury-Brown, a radio astronomer at Jodrell Bank. Hanbury-Brown had been examining the angular structures of radio sources with the 250-foot radio telescope at Jodrell Bank, as these measurements could eventually lead to testing theoretical models of the universe.

While describing his studies Hanbury-Brown drew the attention of his visitors to a special class of radio sources that were significantly smaller than the rest. 'These little chaps' as he called them, stood apart from the typical extended radio sources. Most of them were part of the radio survey that Ryle and his collaborators had made. Did they represent a new class of radio sources? Fred Hoyle felt that they were important enough to deserve special attention not only by radio astronomers but also by optical astronomers.

However, to get an optical identification of a radio source, especially because it was so compact, it was necessary for the radio astronomer to give its position in the sky much more accurately than had been possible till then. Cyril Hazard, then at Jodrell Bank drew their attention to a particular compact source with the catalogue number 3C 273 (the 273rd source in the 3rd Cambridge Catalogue of radio sources). Hazard noticed that the source had low enough latitude so that it could be observed by the Australians too. Then an especially happy circumstance intervened to make the Australian participation not only worthwhile but essential.

The source 3C 273 was due for occultation by the Moon. When an astronomical source like a star or a planet is occulted by the Moon, the Moon's disc covers the source for a brief period while the Moon comes between the source and the planet. Radio telescopes as well as optical telescopes can observe occultations. The advantage is that there is a clear drop in the intensity of the source during an occultation, which picks up again after it is over and these downs and ups can be accurately timed. This information, coupled with astronomers' accurate knowledge of the Moon's position, provides a very accurate position of the source. Thus here was a chance to measure the position of 3C 273 accurately. And the facilities at Parkes were well-suited to the observation.

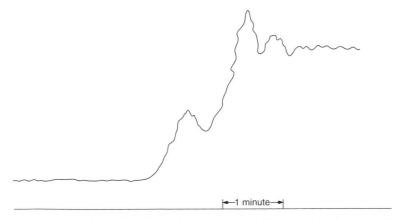

Fig. 10.5. Figure based on the 1962 observations of Hazard, Mackay and Shimmins measuring the drop in radio intensity of the source 3C 273, as it was occulted by the Moon. The drop and subsequent pick up as the Moon crossed over were noticeable. This observation enabled the observers to accurately determine the position of the source in the sky.

This measurement was in due course undertaken by Hazard, Mackay and Shimmins at Sydney who did the measurements at Parkes. The observation went well and the data sets, in duplicate, were flown on two separate flights (to avoid the remote possibility of a plane crash destroying valuable data!) to Sydney for analysis. This gave an accurate position of the source which was then sent to the optical astronomers at Palomar. Was the source a star in the Galaxy or a distant object?

At the radio position of 3C 273 a fairly bright optical star was detected. In the same period when accurate positions of several other radio sources were studied by Allan Sandage and Tom Matthews at Caltech, other, fainter stars appeared. However, when optical spectra of these stars were obtained it was obvious that they were not galactic stars, since they showed a few emission lines at strange wavelengths, a weak blue continuum. What were they? The breakthrough came when Maarten Schmidt and Beverley Oke could identify in 3C 273 a series of emission lines due to hydrogen – the Balmer lines – not at their laboratory wavelengths but with a large redshift $z = 0.158$. Almost immediately another source 3C 48 with $z = 0.36$ was found. With such large redshifts, assuming that they were similar to those found in galaxies, these 'stars' were really very bright objects very far away. They were called 'quasars' or quasi-stellar objects (QSOs).

We mentioned earlier that with large telescopes we rarely discover what they were built to discover, but we find something else. The Hale telescope has contributed many discoveries but none as important as quasars. Because of their large redshifts and large brightness as well as other features compared with normal galaxies the quasars have held out hopes of probing the past history of the universe. However, there is a snag! As we will describe in Chapter 16, there are serious doubts in the minds of some astronomers as to whether or not the redshifts of quasars arise entirely from the Hubble effect of expansion of the universe.

11

The origin of the chemical elements

The base particles

Nearly all of the matter and energy in the universe that we observe is contained in the form of protons (hydrogen nuclei), neutrons, electrons and their positively charged mirror particles, positrons, together with neutrinos and the antiparticles, and photons (radiation) which are particles of the electromagnetic field. The neutrons and protons are the heaviest of these particles and are called 'baryons' (from the Greek word *barys*, meaning 'heavy'). This is all the basic 'stuff' which must have been created somehow, whatever kind of a universe we believe in. This creation also includes the creation of the laws of physics.

However, what we see all around us on the Earth, the planets, and the stars, and what we are made of ourselves is something rather different. We see and can detect the photons of all energies (wavelengths), and we can detect particles, with very high energies which move very fast – they are called 'primary cosmic rays' and are mostly protons and electrons moving at speeds close to the speed of light. They pour into the Earth from all directions. And then there are the neutrinos – particles which have practically no mass, moving at the speed of light. They are found everywhere, but they are very difficult to detect because they interact with the rest of the matter very feebly.

But much of the matter – the 'stuff' around is made up of chemical elements which are much heavier than individual protons or neutrons. The relative numbers of the chemical elements from the lightest, hydrogen, to the heaviest, and how they were made is the subject of this chapter.

Since each chemical element consists of a nucleus which contains many protons and neutrons surrounded by a cloud of electrons, it is reasonable to suppose that the elements are the result of building processes, each step being the addition of a single proton or neutron. Where and how did this take place?

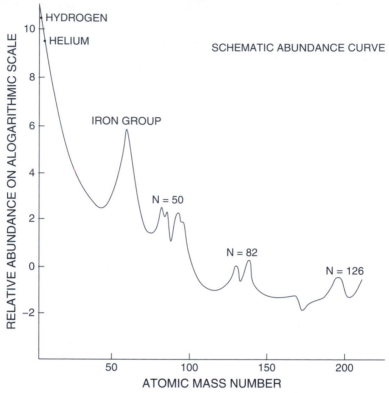

Fig. 11.1. The relative abundances of chemical elements are shown. The peaks indicate high abundances. As we move to the right we encounter elements of larger atomic mass. N here in the charge Z defined in the text.

Building the elements

The relative abundances of these chemical elements are shown in the figure.

The weight of each element A is measured in atomic units, each unit being equal to the mass of a proton, the nucleus of a hydrogen atom $= 1.6 \times 10^{-24}$ grams.[1] This is to be compared with the mass of the electron $m_e = 9 \times 10^{-28}$ grams. The electron is much lighter than the proton – the ratio of their masses is 1 to 1836.

Each nucleus has a positive electric charge Z where the unit of charge $= 4.8 \times 10^{-10}$ electrostatic units. It is customary to write a nucleus X containing A particles and charge Z as $^A_Z X$. We will drop the suffix Z in our text to follow.

All nuclei have positive charge $(+)$ which corresponds to the number of protons in the nucleus. This positive charge is balanced by the electrons around it, since each electron has a negative charge $(-)$. Thus the net charge

---

[1] 10^{-24} is a compact way of writing 1 divided by the very large number obtained by writing 24 zeros after 1. *see* Chapter 13.

of the atom is zero, so it is *neutral*. Thus for a helium atom, the nucleus has $A = 4$, and $Z = 2$. (A neutron has no charge, i.e. $Z = 0$.) This means that the helium nucleus contains two protons, each of which has a positive charge, and two neutrons each of which has zero charge. Since the helium atom has two electrons, the total charge $= +2 - 2 = 0$.

As can be seen from the figure, hydrogen is the most common chemical element. It is followed by helium, then the very rare elements lithium, beryllium and boron and then carbon $^{12}C(A = 12, Z = 6)$, nitrogen $^{14}N(A = 14, Z = 7)$, and oxygen $^{16}O(A = 16, Z = 8)$. In each case we have only named the most abundant *isotopes* of these elements. Many elements have several *isotopes* – these are nuclei with the same charge Z, but different atomic weights (A). For example, most of the hydrogen in the universe has $A = 1, Z = 1$, but there is a rare isotope of hydrogen often called deuterium, or ^2H with $A = 2$, $Z = 1$. Also, most of the helium is ^4He but there is an isotope ^3He with $A = 3$, $Z = 2$. In the first case the ratio of deuterium, sometimes called heavy hydrogen, to hydrogen is 1 to 10 000.

While the abundances of carbon, nitrogen and oxygen are very small compared with hydrogen, the abundance curve for higher values of the atomic weight A falls precipitously, and then has a tendency to level off in the region of $A = 50$–60 with a minor peak in the region of iron, $^{56}Fe(A = 56, Z = 26)$. The abundance curve then falls further as we reach the heaviest elements, in the region of lead, ^{208}Pb, and uranium, ^{235}U and ^{238}U.

This general picture of element abundances was put together by astronomers over many decades by studying the compositions of stars, particularly the Sun, the Earth, and the meteorites.

Most of the abundances were obtained from spectrum analysis of the Sun (*see* Chapter 6). As atomic physics was developed in the first decades of the twentieth century, spectroscopists showed that each chemical element on the Earth when heated, and examined with a spectrograph gave rise to a unique spectrum, each line arising from a transition between energy levels of the electrons surrounding the nucleus of that atom. Thus each element has a characteristic signature which after much effort was identified in the spectrum of the Sun in which all of the elements are mixed together.

The Earth and the meteorites are part of the Solar System made up of planets and other fragments which formed at the time that the Sun was first becoming a star some four and a half billion years ago. In the process of formation of the planets, most of the mass in the form of the most common elements, hydrogen and helium, was removed. But the heavier elements which have many isotopes remained. We are able to obtain the abundances of the isotopes of the heavier elements by studying samples from rocks and meteorites.

Until about 1925 astronomers assumed that the Sun was mainly composed of heavy elements like iron. But a young English graduate student, Cecilia Payne, first working in Cambridge in England, and then at Harvard, in Cambridge, Massachusetts, did the first spectrum analysis of the Sun. She was the first to use the new-fangled quantum mechanical methods. In her thesis, published in 1925, she reached the startling conclusion that the most abundant element was hydrogen and not iron. For some years this result was not accepted by the two leading theorists in astrophysics, Sir Arthur Eddington in Cambridge, England and Henry Norris Russell at Princeton in the USA. They both believed in the models they had made of the Sun in which iron was

Fig. 11.2. Cecilia Payne-Gaposchkin (1900–1980). Credit: *Pencil sketch Anagha Pujari.*

assumed to dominate. In the latter part of her work, Cecilia Payne was supported by Russell, who had secured her a position at Harvard, and in her published thesis she was apparently so embarrassed by this result, that she stated in the conclusion that she realized that the result must be wrong, but she could not see how.

But of course she was right. She remained at Harvard for the rest of her life and ultimately was made Professor and Head of the department, though for decades she was discriminated against. Russell was finally quoted as having said that her Ph.D. thesis was the best that had ever been written on astronomy. This after he had disbelieved the result for two or three years.

To build the heavier elements from protons, neutrons and radiation, it is necessary to find an environment when the density and temperature are high enough so that nuclear reactions can occur. One obvious place where this can take place is in the central regions of stars.

Just before World War II, in 1938, Hans Bethe and others had solved the basic problem of how all of the energy was generated in the stars to make them shine. The process is one in which at the centre of the star it is hot enough and dense enough for a set of nuclear reactions to lead to the conversion of hydrogen to helium. We call it the burning of hydrogen to form helium. This requires processes in which 4 protons are converted into one helium nucleus (sometimes called alpha particles by nuclear physicists). Since the mass of the 4 protons is greater than the mass of one helium nucleus, the difference is released as energy. Using the Einstein relation $E = Mc^2$, this tells us that the conversion of 1 gram of hydrogen to helium releases up to six times a million million million ergs (6×10^{18}). This is what makes the stars shine.

There are two types of nuclear reactions which can do this. These are shown in the box.

Nuclear fusion of hydrogen

The nuclear reactions leading to the formation of helium from hydrogen can proceed in stars in two ways.

The p–p chain:

The nucleus of hydrogen is just the proton p. The fusion of four protons to make a helium nucleus can proceed as follows. First two protons combine to make a nucleus of heavy hydrogen or deuterium d. A positron e^+ (the antimatter counterpart of the electron with positive charge) and a neutrino ν are also produced. The neutrino has no charge and very little mass, if any.

$$p + p \rightarrow d + e^+ + \nu.$$

The deuterium then combines with a proton to produce an isotope of helium of mass 3 along with radiation (denoted by the symbol γ):

$$d + p \rightarrow {}^3\text{He} + \gamma.$$

Finally two such light isotopes of helium combine to make one normal nucleus of helium while releasing two protons:

$${}^3\text{He} + {}^3\text{He} \rightarrow {}^4\text{He} + p + p.$$

Thus in the end we have the helium nucleus in place of the four hydrogen nuclei. This process is believed to take place in the Sun.

The C–N–O cycle:

This is a cyclic series of reactions that uses nuclei of carbon, oxygen and nitrogen to facilitate the fusion process. These nuclei act as catalysts. The cycle may start at any stage, but we begin with carbon.

$${}^{12}\text{C} + p \rightarrow {}^{13}\text{N} + \gamma.$$

Thus the normal carbon nucleus combines with a proton to make a light isotope of nitrogen, which decays into a heavy isotope of carbon:

$${}^{13}\text{N} \rightarrow {}^{13}\text{C} + e^+.$$

This isotope of carbon then picks up another proton to change into a normal nucleus of nitrogen; some radiation is also emitted in the process:

$${}^{13}\text{C} + p \rightarrow {}^{14}\text{N} + \gamma.$$

The nitrogen nucleus then collects another proton to make a light isotope of oxygen, also emitting radiation:

$${}^{14}\text{N} + p \rightarrow {}^{15}\text{O} + \gamma.$$

Next, this isotope of oxygen decays into a heavy isotope of nitrogen, of mass 15 while producing a positron also:

$${}^{15}\text{O} \rightarrow {}^{15}\text{N} + e^+.$$

The final step in the cycle is for the nitrogen isotope to combine with a proton to produce a helium nucleus and also a normal nucleus of carbon with which our cycle had started.

$${}^{15}\text{N} + p \rightarrow {}^{12}\text{C} + {}^4\text{He}.$$

The CNO cycle usually acts in stars of high mass, say mass exceeding that of the Sun.

If we only have protons and neutrons to start with, then helium can only be built through the proton–proton chain. This is the main source of energy production in the stars. The central temperature of the Sun where this goes on is about 13 000 000 degrees Kelvin and the density is about 100 gram per cm^3. If carbon and nitrogen are already present in the matter out of which the star is made and the star's centre is hotter, then the CNO cycle will be able to produce the helium.

By following the CNO cycle through in the box it is clear that the carbon, nitrogen and oxygen are not being used up. They simply act as what are called *catalysts* to enable four protons to be converted to one helium nucleus. Their role is to facilitate the nuclear processes.

Enter George Gamow

This was the state of knowledge after World War II. By about 1947, George Gamow, a very talented theoretical physicist from Russia, who had lived in the West for many years, was fascinated by cosmology and began to study the big-bang universe and the element abundances. The first problem was the origin of helium. George knew that it could be built by hydrogen burning in the centres of stars, but its abundance is so high (about 24% of the total mass) that he concluded that there had not been enough time, since the beginning, to build that much helium. In those days it was thought that the 'age' of the universe could not be greater than about 2 thousand million years. This was based on the rate of expansion which had been derived by Hubble and Humason (*see* Chapter 10). The 'age' is inversely proportional to the value of the Hubble constant. Gamow could see that in 2 billion years only a small fraction of the helium could have been made. Thus he and a number of other leading physicists of that time concluded that the helium must have been made in the first few minutes of the big bang, when they assumed that the matter would be in the form of a dense cloud of protons, neutrons, electrons and radiation.

There was no theory that could predict what the ratio of the number of baryons (the protons and neutrons) to the number of photons (radiation) would be in the beginning, but if the right choice was made, in the first minutes in the history of the universe, the reactions would go on to build enough helium to explain what we can see, plus some deuterium (heavy hydrogen) ^{3}He and a trace of lithium in the form of ^{7}Li.

In the scenario of primordial nucleosynthesis, the universe cooled from around 10 billion degrees to a few hundred million degrees as it aged from 1 second to 200 seconds, or a little over 3 minutes. As it cooled, its constituent

Fig. 11.3. George Gamow (1904–1968).
Department of Physics, Texas A&M University – Commerce.

baryons (neutrons and protons) slowed down. Now as any two of them came close together their attractive nuclear force would come into play (this force acts only at a short range of around a million-billionth part of a metre). At sufficiently slow speed a neutron and a proton would be trapped together to form a *deuteron* or a nucleus of heavy hydrogen. Further such reactions would continue and deuterons in turn would combine with more baryons or deuterons to eventually form the nucleus of helium.

All this must happen in those first three minutes. As mentioned before, this *can* happen, provided the right density of baryons exists at the right temperature. For example, if the number density of baryons exceeds a critical limit, too many deuterons would form intially; and with so many of them around they will all take part in further nuclear binding and be transformed into helium. So if one wants to explain how the deuterons seen today came about in Gamow's primordial nucleosynthesis process, then one is forced to stipulate that the baryon density in the universe today should be *less than* about 4% of the critical density of matter defined in Chapter 7. This assumption must be made to proceed further with Gamow's programme of making heavier nuclei.

However, when attempts were made to see if heavier and heavier elements could be built to explain the abundance in the figure, they all failed, because in nature there are no stable elements with mass $A = 5$, or $A = 8$. Thus by 1950 it had become clear that if there was a big bang, the lightest elements, D, He and some Li could have been made, but not the heavier elements.

Therefore they must have been made somewhere else.

Fred Hoyle's breakthrough

The scientist who cracked that puzzle was again Fred Hoyle. For some years, he had been studying the abundance of the elements, and he noticed that in the region of iron (^{26}Fe) there tended to be a relatively smooth, broad, but small peak in the relative numbers of nuclei on each side of the peak.

Now it is known that iron is the most stable nucleus in the periodic table. It takes more energy from the bombarding particles to break it apart than the surrounding nuclei. The shape of the curve suggested to Hoyle that these nuclei must have been made in what physicists call an equilibrium situation in which all of the nuclei are colliding with each other in a region of very high density and temperature, and they spend enough time together to reach an equilibrium.

In a paper that Hoyle published in 1946, two years before he developed the steady-state cosmology, he concluded that the iron peak elements must have been produced in an equilibrium situation in which the temperature was about three and a half thousand million degrees (3.5×10^9 degrees). He concluded that this situation could only have taken place at some time in the evolution of the stars. Hoyle had made a real breakthrough. He had proposed that elements were being made in stars in the course of their evolution.

But how could the elements heavier than helium have been cooked inside the stars? We already know that for most of their lives stars exist and shine by burning hydrogen in their deep interiors, and that the temperatures required lie in the range of about 10 to 50 million degrees, far less than the temperature required to lead to reactions of the type suggested by Hoyle. To understand how higher temperatures are reached we have to discuss the work that was done in the 1950s and 1960s which led to an understanding of how stars evolve from birth to death.

After stars form by collapsing from low density clouds of gas, their centres get hotter and hotter until the temperature is high enough so that collisions between the nuclei lead to the building of helium from hydrogen. They then stop collapsing and are stable for a long time. At this stage how big they are and how brightly they shine depends on two properties – their mass – meaning how heavy they are; and what they are made of. The most important quantity is their mass. The larger the mass, the brighter they are.

Thus if we plot the brightness – the luminosity as it is called – against the temperature at the surface in the star's atmosphere we find that all stars, for most of their lives, lie on a line called 'the main sequence' shown in Figure 11.4.

Here, we have marked the position of the Sun. It has a mass of 2 times a million-million-million-million-million kilograms (2×10^{30} kg) and a surface temperature of about six thousand degrees (6000 K). It turns out to be useful to define the mass of a star in solar units (1 solar unit = $1 \ M_M$).

Calculations of the rate at which stars burn hydrogen in their centres, leaving helium ashes, have been made for stars of all masses between a fraction of a solar unit, and about 100 solar units.

After a fraction of the mass at the centre of a star has been burned, its structure begins to change, since there is no longer a source of nuclear energy at the centre. It turns out that after about 12% of the mass is burned, the inner parts of the star will tend to fall in since there is no energy source except gravity available to build it there. However as it falls in it gets hotter, and the regions around the centre will burn hydrogen, while in the very centre temperatures as high as two hundred million degrees will be reached.

At that temperature the helium nuclei are moving so fast that when they collide with each other two of them can stick together to form beryllium eight (^{8}Be), which is an unstable nucleus, which if left to itself will decay. But, in some cases before this can happen this nucleus will collide with another helium nucleus to form carbon twelve (^{12}C). This is a stable nucleus and it means that at high enough temperatures three helium nuclei can form carbon twelve. This process is called helium burning – it was first worked out by Ed Salpeter, an Australian nuclear astrophysicist at Cornell University in the USA.

However, there was one gap in this argument. The chance of three helium nuclei coming together was small. To compensate for this, Fred Hoyle argued

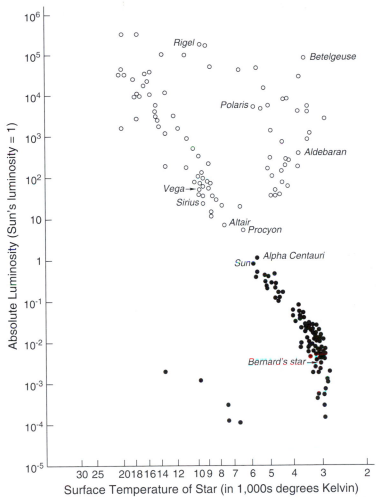

Fig. 11.4. The above figure is called a Hertzsprung–Russell diagram, after the two astronomers of those names. The diagram plots on the horizontal axis, the surface temperature of a star, decreasing from left to right as marked. The vertical axis plots the luminosity of the star, increasing upwards. Notice that when stars are plotted as above, the majority lie on a track extending from the bottom right to top left. This is called the main sequence. The Sun is marked on the main sequence...as we see, it is an average star with average brightness and temperature.

that when they do come together they *react very fast*. This can happen when there is a resonance, i.e., when the energy of ^{12}C created *exactly* matches the energy of the three helium nuclei. Hoyle found that this matching would imply that an 'excited' level of ^{12}C must exist, with energy higher than the normal carbon found in nature. When on a visit to Caltech Hoyle asked the nuclear physicists there, Willy Fowler and Ward Whaling, to search for such a nucleus in their experiments. At first they disbelieved him; but on further persuasion they looked and found it! This cleared the path for synthesis of nuclei heavier than ^{12}C.

This shows that in stars, but not in a big-bang, the temperatures can get hot enough for sufficient time to build the event and get past the regions of mass 5 and 8 in which there are no stable elements.

Towards heavier nuclei

As the star continues to evolve it gets hotter and hotter in its interior, and more nuclear reactions can take place involving nuclei with large charges Z colliding with helium nuclei. Thus a helium nucleus collides with carbon twelve to give oxygen sixteen (^{16}O), and then successively neon twenty (^{20}Ne), magnesium twenty-four (^{24}Mg), silicon twenty-eight (^{28}Si) and so on. As the star continues to get hotter and shrinks, this process of building heavier elements from lighter ones will continue until we reach the most stable nuclei near iron.

Up to this stage, each nuclear reaction leading to a heavier nucleus will have *released* energy. But this stops near iron, and to build all of the heavier elements beyond iron we have to find new sources of energy.

The star up to this point can be seen as a three-dimensional layer cake or as an onion with skins, with the highest temperature layer in the centre, containing iron and outer layers containing progressively lighter nuclei; the outermost containing hydrogen.

But what happens when we reach iron? The only source of energy left in the star is gravity. Thus the central region will collapse, and pull in the matter after it. The only source of energy is now if the iron breaks up again all the way to protons and neutrons. This takes place in only a few seconds, and a tremendous amount of energy will be released, much of it in the form of neutrinos. These neutrinos only react very weakly with neutrons and protons, but it has been surmised that even so, the energy may be great enough to reverse the collapse and blow the outer parts of the stars away. In the course of this a very large number of neutrons will interact with the remaining heavy nuclei, and this process is able to build many of the isotopes much heavier than iron. The process will pile neutrons on to the heavier and heavier elements until an atomic number is reached at about $A = 240$ where the nuclei spontaneously decay because they are so unstable, and we are left at the high end of the abundance table with uranium ($A = 235$, $Z = 92$ and $A = 238$, $Z = 92$), both of which will also decay but which are present on the Earth (and have recently been found in stars).

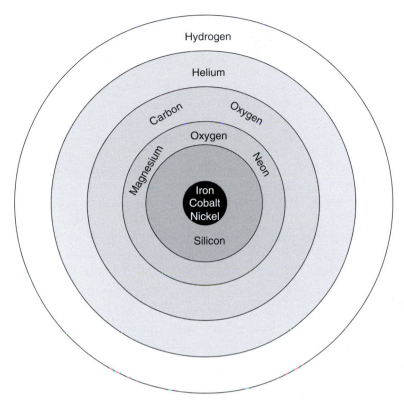

Fig. 11.5. The red giant star after considerable evolution develops layers within, each layer of different chemical element. Starting with the elements of the iron group at the centre the layers progressively contain lighter elements, with hydrogen in the outer envelope.

It is also in these few seconds that the equilibrium process originally suggested by Hoyle occurs. One of us (Geoffrey Burbidge) was part of a 4-member team consisting of Margaret and Geoff Burbidge, Willy Fowler and Fred Hoyle, who in 1957 did the first comprehensive work covering all aspects of stellar nucleosynthesis. Since the work of 1957 many very elaborate calculations have been made, using hundreds of nuclear reactions and using very large computers. We are still trying to understand the fine detail of what exactly occurs.

There is strong evidence that the explosions of the stars suggested by theory are the supernovae which were first identified by Walter Baade and Fritz Zwicky at Mount Wilson in the 1930s. They suddenly appear in distant galaxies and become very bright and then the light decays in a few weeks.

The remnants of the star left in the middle will continue to collapse. If the mass is less than about 2 solar units it will ultimately become stable, but its

Fig. 11.6. The Crab Nebula, deriving its name from the filamentary structure of the ejecta after the explosion of a star. The explosion was observed and recorded by Chinese observers. The date of observing the explosion can therefore be given precisely as July 4, 1054. What we see today some nine and a half centuries later is the debris of the explosion still shining in the form of the above nebula. Credit: *NASA*.

density will be as great as that of a single nucleus. It will then be a *neutron star* with a size of only about 10 kilometres (about as big as a major city) but the total mass of the Sun. If it has a mass much greater than this, nothing will stop it from shrinking indefinitely and it will become a *black hole*. The theory behind this was worked out by Subrahmanyan Chandrasekhar (Chandra) when he was a very young man in the 1930s. He showed that for masses less than about 1.4 solar units, a star can end up as a white dwarf star, when its density is a million times that of water. The neutron star is an even more compact state of a star. Chandra received a Nobel Prize for this work in 1983.

There is much more that we could say about supernovae, neutron stars and black holes, but here we are concentrating on the formation of the elements.

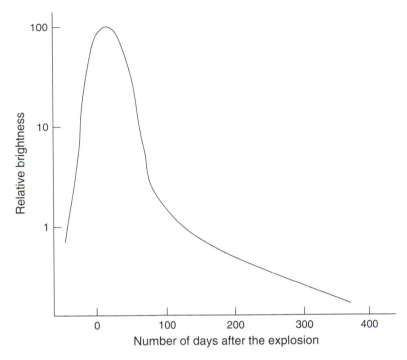

Fig. 11.7. A typical light curve of a supernova showing the rapid rise and slow fall of light emanating from the exploding star. Notice that the slow fall lasts over several weeks.

Stellar evolution

Earlier we explained that a star spends much of its life on the main sequence burning hydrogen, but that after a fraction of the mass is burned the structure will change.

For the Sun it will take about 10 billion years for this to occur (the age of the Sun is now about 4.6 billion years). But for stars with larger masses it will take much less time for the same thing to occur. For a star of perhaps 50 solar units, it will only take about a million years for the structure to change. This is because a heavier star burns much more rapidly than a lighter one and so exhausts its fuel faster.

Thus it follows that all of the stars we see around us on the main sequence have very different ages. The massive stars are very young, and stars like the Sun are very old.

What happens when the structure changes? It turns out that stars with masses close to that of the Sun get slightly cooler on their surfaces, and move to the right in Figure 11.8. The calculations show that they then tend to get

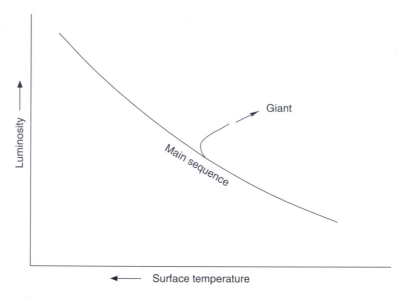

Fig. 11.8. This diagram shows with the help of arrows the direction from the main sequence in which a star evolves.

brighter as they are getting bigger – so that they move into a part of the diagram called the red giant region. Here their energy sources are helium burning at the very centre, surrounded by a hydrogen burning zone.

Detailed calculations show that the red giants only spend a fraction of the time they live on the main sequence, on the red giant branch. After that they move back in the diagram towards the main sequence, but on the way they may begin to engage in regular pulsations becoming regular variable stars.

As we said earlier, some, depending on their mass, may explode and become supernovae. Others have phases in which they eject matter at high speeds but over millions of years. These are often seen as planetary nebulae, and they may quickly end their lives as small, very dense stars which cool slowly. They are the white dwarfs. They are of a size similar to the Earth, but their masses are also very close to that of the Sun.

Stars in the red giant stage contain many elements including hydrogen, carbon, and helium. We have described the building of elements by successively adding helium nuclei, but it is also the case that nuclear reactions involving odd isotopes (for example $^{13}C + {}^{4}He$) will take place and this will lead to free neutrons (as in $^{16}O + n$). While they decay in a few minutes they may first be captured by other nuclei thus giving rise to some of the neutron-rich isotopes. It is through processes like this that many of the isotopes with odd numbers are built.

Fig. 11.9. This ring-shaped nebula is a planetary nebula (called the *Ring Nebula*) in which the ejection of material from the star is relatively modest compared to that of a supernova like the one in the Crab Nebula. Credit: *NASA*.

We have direct evidence for this in some red giant stars where ^{12}C and ^{13}C have been detected in the spectra. The most spectacular case was the discovery of technicium 43 (^{43}Tc) in a giant star by Paul Merrill. This element is not stable. It decays spontaneously in a few million years. Thus its detection in an old star is proof that nuclear processes involving neutrons are going on now.

We have now shown how the main processes giving rise to all of the elements with the possible exceptions of helium and deuterium, take place in stars. Both neutrons added slowly over a long time scale in stars like red giants and neutrons added very rapidly in the last few moments of supernovae explosions are thought to occur. The slow process has been called the *s*-process, and the rapid process, the *r*-process.

The elements are part of the gas which is ejected from the stars as they evolve. They form clouds out of which new generations of stars are formed. What is left behind are white dwarfs, neutron stars and black holes.

The working out of this whole complicated process first started with the ideas of Fred Hoyle. In the 1950s Al Cameron, a Canadian nuclear physicist (1925–2005) worked out much of the detail. Then came the group involving Fred Hoyle, Willy Fowler (an experimental nuclear physicist from Caltech),

Fig. 11.10. Margaret Burbidge, Geoffrey Burbidge, William Fowler and Fred Hoyle, the four astronomers who are credited with the comprehensive work on nucleosynthesis in stars. Here they are admiring a model railway engine presented to Willy Fowler on his sixtieth birthday. Credit: *Fred Hoyle, private collection.*

Margaret Burbidge (an observational astronomer) and Geoffrey Burbidge. This work was published in 1957 in *Reviews of Modern Physics* and is commonly referred to as the paper by B^2FH.

Stellar nucleosynthesis is now generally accepted as the way that the elements were formed. Thus whether or not it ever occurred, the big bang did not give rise to the chemical elements. Stars made almost everything, and the reader should be aware that everything he or she is made of was once inside stars.

12

Cosmic microwave background

Introduction

The most obvious and most easily detected feature of the universe is that it is made up of immense numbers of isolated lumps of matter which are radiating, and the lumps are all expanding away from each other. These lumps – galaxies – which are the fundamental building blocks of the universe, with a range of morphological types, are sometimes connected in groups and clusters. We assume that these are held together by gravitational attraction.

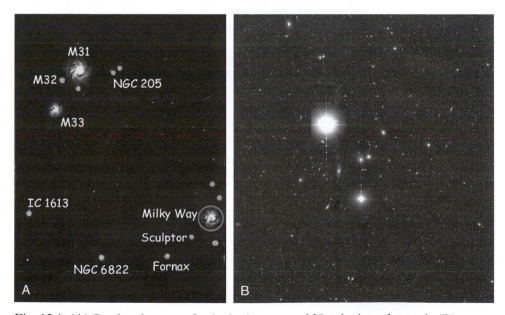

Fig. 12.1. (A) Our local group of galaxies has around 20 galaxies, whereas in (B) we see a cluster which has hundreds of them. Credit: *A) Arvind Paranjpye. B) Atlas Image courtesy of 2MASS/UMass/IPAC-Caltech/NASA/NSF.*

Table 12.1

Type of radiation	Wavelength λ, Frequency ν, Energy range E	Energy density (erg/cm^3)
Radio	$\nu \leqslant 4080$ MHz	$\leqslant 10^{-18}$
Microwaves	λ in 80 cm to 1 mm	$\sim 4 \times 10^{-13}$
Optical	λ in 4000−8000 Å	$\sim 3.5 \times 10^{-15}$
X-rays	E in 1−40 keV	$\sim 10^{-16}$
γ-rays	$E \geqslant 100$ MeV	$\leqslant 2 \times 10^{-17}$

If you happen to live in a big city, you will have discovered that the night sky is not totally dark. It has a yellowish glow, which in fact arises from the light emitted by the various urban sources of light, by street lamps, the neon lights of advertising, lights from houses and apartments and, of course, headlights of passing vehicles. The light from these sources gets scattered many times by dust particles, reflected back by clouds above and manages to be spread around. This is the diffuse glow you see in the sky, which is of course a *bête noire* for astronomers who cannot see their faint stars and galaxies against the background of this glow. Also, because of multiple scattering you cannot make out from which individual sources a particular part of the glow arises.

Since the galaxies are all radiating energy over a wide spectral range we expect that there exist diffuse radiation fields of starlight, infrared energy, radio energy, X-ray energy, etc. Like the glow of the urban night sky, one sees overall radiation without being able to identify the original sources of energy. Over many years it has been possible, however, to estimate energy density in the different energy bands. The results are shown in Table 12.1. We may mention that the unit of energy used by astrophysicists is the 'erg' which is twice the energy of motion of a particle of 1 gram of matter moving at the speed of one centimetre per second. An expression like 10^{-18} means the tiny fraction obtained by dividing 1 by the large number 1 000 000 000 000 000 000, having 18 zeros after 1.

Other diffuse energy components are the magnetic energy, and the cosmic ray energy, the latter being largely carried by fast-moving protons with very large energies. Finally and most importantly there is direct evidence for energy radiated mainly in the microwaves. This has by far the largest energy density as seen from Table 12.1.

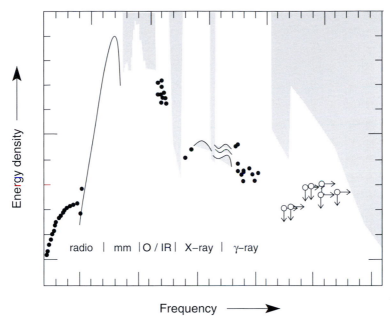

Fig. 12.2. A schematic plot of the energy of cosmic radiation at different wavelengths. It is obvious that the maximum energy resides in microwaves (the millimetre range).

Black-body radiation

This 'champion' amongst all energy densities, the so-called *microwave background*, is in the form known as 'black-body radiation'. What is a black body? What does one mean by radiation from a black body? A simple example from household experience will help understand these questions. Imagine a person baking a cake in an oven. Before placing the dough in the oven they heat it by setting a specified temperature, say 250 degrees Celsius on the dial. This action activates the heating elements in the oven. They get hotter and hotter and radiate heat. This heat is absorbed by other parts of the oven and they start warming up. All along the hotter parts radiate heat to warm up cooler parts. This process goes on until all parts in the oven are at the same preset temperature. Then the heating is switched off by the oven's thermostat. From then on the oven, ideally maintains this temperature *everywhere* in its interior.

We say that such an oven has reached the state of a *black body*. In such a state, no heat escapes outwards and the trapped heat has reached a state of equilibrium when all parts of the interior have the same temperature. Each part radiates as much heat as it receives. The body is black because no radiation escapes from it.

A black body therefore has in its interior radiation travelling back and forth. This radiation is in the form of electromagnetic waves. These waves

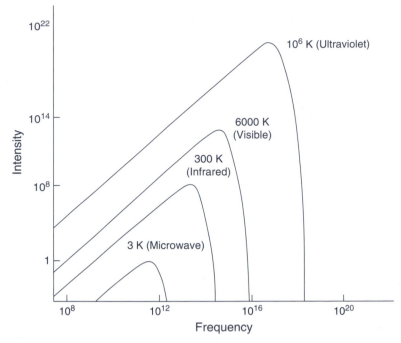

Fig. 12.3. This figure shows how the energy of black-body radiation is distributed at different wavelengths. The typical curve peaks at an intermediate wavelength, dropping to low levels as we move away from it. The different curves indicate how they differ according to the temperature of the black body. As the temperature increases, the curve rises and its peak shifts towards shorter wavelength and higher frequency.

come in a range of wavelengths. Typically the emitted radiation is very small at very long and very short wavelengths but peaks at some intermediate value. The value at which it peaks depends on the temperature of the radiation. The German physicist Wilhelm Wien discovered this relationship through experiments and his result is known as *Wien's law*. Simply stated it is that the peak wavelength multiplied by the temperature (expressed in absolute units) gives the same figure for all black bodies.

The absolute scale of temperature has a zero at −273°C and each unit of temperature equals 1°C. Thus 100 on the absolute scale means −173 on the Celsius scale. The Wien law tells us that a black-body radiation at 100 absolute temperature will peak at wavelength 0.0029 centimetres, while a black-body radiation at 1000 absolute temperature will peak at a wavelength ten times shorter, that is at 0.000 29 centimetres. The total energy density of black-body radiation rises in proportion to the fourth power of its temperature.

So the important thing to notice about the microwave background radiation is that its spectrum follows the same pattern as that of a black body of temperature 2.73 on the absolute scale.

Origin of various backgrounds

It is important to appreciate what led astronomers to an understanding of why most of these diffuse energy fields are present. Since all electromagnetic energy can freely propagate from its sources, it was easy to understand the background due to starlight. Similarly the infrared flux originates from dusty galaxies, and from cool stars in these galaxies, and from any regions in which optical energy is scattered and re-radiated in the infrared before it leaves the galaxy. In the last few years surveys made with X-ray telescopes have shown that this background radiation is made up of large numbers of discrete X-ray sources, active galactic nuclei and X-ray emitting QSOs. It looks as though at least 70% of the background in this energy range can be explained in this way. Whether or not all of the background X-rays can be explained in this way is not known. Similar arguments based on extensive surveys of extragalactic radio sources lead to the estimate of background radio emission. For the other estimates we limit ourselves to our own Galaxy – the Milky Way.

Detection of magnetic fields in spiral arms in the Galaxy is very difficult, and it has only been accomplished in our own Galaxy where fields of the order of $B \simeq 5 \times 10^{-6}$ Gauss are found corresponding to to an energy density $\simeq 10^{-12}$ erg/cm^3. The value is higher in dense molecular clouds and the average value over our Galaxy is not known.

As far as the cosmic ray energy density is concerned we are dealing here with charged particles that cannot escape freely from our Galaxy where it is believed most of these were generated and which may be destroyed by collisions with interstellar atoms or molecules, or they may escape. Their energy density which amounts to about 5×10^{-13} erg/cm^3 is determined by a balance between production (in supernovae or pulsars) and escape or annihilation.

The most important radiation field first detected in our Galaxy is the microwave background which has a black-body temperature close to 3 K and consequently with an energy density $\simeq 4.5 \times 10^{-13}$ erg/cm^3, is much more energetic than the other energy fields listed in Table 12.1.

This radiation field was first detected in 1941 through its effect on interstellar molecules seen in the spectrum of stars in the Galaxy by A. McKellar. The molecules of cyanogen were identified as CN and CN$^+$, and from the lines that were detected and limits on others it was shown by McKellar (1941) that a black-body radiation field must be pervading the whole Galaxy. Its temperature T was set by McKellar to lie in the range $1.8 < T < 3.4$ K.

This was a very clever way of detecting the radiation. For atomic physics tells us precisely how ambient radiation can excite atoms and molecules. Finding some molecules (of CN and CN$^+$) in excited states enabled McKellar to make a crude but definite estimate of the temperature of the background radiation.

Of course at the time there was no identification of the source of the radiation nor any knowledge as to whether or not it was purely galactic in origin. In this sense, the microwave background seemed to stand apart from other radiations we have mentioned here.

The microwave background and the big bang

When George Gamow and others began to speculate on cosmology knowing that the universe is expanding, they concluded that it was likely that at an earlier epoch when all of the matter was very much closer together, nuclear processes might have built the heavier elements. One side effect of this work led Gamow's younger colleagues Ralph Alpher and Robert Herman to the important conclusion that, apart from nuclear particles, the developments in the early universe would also leave as relics, photons, particles of radiation. Alpher and Herman made an inspired guess that the relic radiation (which would be of black-body character) would have a present day temperature of

Fig. 12.4. Robert H. Dicke and David Wilkinson, pioneers in the building of detectors for measuring the cosmic microwave background. The Microwave Anisotropy Probe (MAP) satellite launched in 2001 was named after Wilkinson and became known as WMAP. Credit: *Department of Physics, Princeton University*.

Fig. 12.5. Arno Penzias and Robert Wilson in front of their antenna which measured cosmic microwave background radiation, in 1965. Credit: *Alcatel-Lucent*.

around 5 degrees on the absolute scale. At that time they had no knowledge of the measurements by McKellar showing the existence of the microwave background.

As we discussed in Chapter 11 it is now established that only ^{2}D, ^{3}He, ^{4}He and ^{7}Li could be made in a big bang. However, Gamow and his colleagues were very much attracted to the fireball model and showed that as the universe expanded, its black-body nature would be retained with steadily diminishing temperature.

Robert Dicke and the Princeton group took up this idea and tried to test it by detecting the radiation; an experiment was being set up in Princeton by Dicke and his colleague David Wilkinson. However, in an unexpected way they were deprived of being the first to measure the radiation directly by using radio antennae.

This happened because, unknown to them, two scientists, Arno Penzias and Robert Wilson, in the Bell Telephone Laboratory at Murray Hill, New Jersey were engaged in some measurements of what they thought would be man-made radiation at 7.3 centimetres wavelength. For their study they had set up a horn-shaped antenna.

To test and debug their antenna, Penzias and Wilson took sample measurements in different directions in the sky. When they eliminated the contributions from the various known sources, they were surprised to find a residual radiation that was left unexplained. Moreover, the radiation appeared isotropic; that is, the same in all directions. Penzias and Wilson were unaware of the theoretical work of Gamow and his colleagues, or of the Princeton group of Dicke *et al.*, let alone the pioneering work of McKellar. Their first suspicion (natural in these circumstances) was that there was some unknown bug in their antenna. They even got rid of pigeon droppings on the horn just in case it gave the apparently spurious results!

However, in 1965, the scientific grapevine put the Princeton group in touch with them and when they relooked at their unexplained radiation they could interpret it as relic radiation with a black-body temperature of around 3.5 K. Here K denotes the absolute scale of temperature named after the nineteenth century scientist Lord Kelvin. However, it was not until 1990 that J. C. Mather and others confirmed that this radiation very accurately followed the black-body spectrum for a temperature of 2.726 K.

This is obviously the radiation indirectly detected by McKellar back in 1941. Many observations have by now shown that it is not confined to our Galaxy, but is truly of extragalactic origin. Did it arise in a big bang as nearly everyone currently believes?

We have shown earlier that most of the other diffuse radiation fields are made up of radiation from many discrete sources. However, the early work and ideas of Gamow clearly led most cosmologists to believe that, with the discovery by Penzias and Wilson, the problem of the origin of the microwave background had been solved.

The history of the subject and particularly the fact that none of the cosmologists knew of McKellar's work has led to the erroneous idea that Gamow made a prediction and Penzias and Wilson showed it to be true. In the more than forty years which have elapsed since the Penzias–Wilson result, more and more sophisticated observations have been made of microwave background radiation. One of these referred to earlier was by Mather and his colleagues who used COBE (Cosmic Background Explorer) satellite-based measurements and showed that the radiation has a rather precise black-body form with temperature 2.736 K, well within the limits originally predicted by McKellar.

In this connection, it is worth highlighting the fact that the basic physics of big-bang cosmology does not lead to a prediction of what its present day temperature should be. Alpher and Herman came up with the figure of 5 K. Gamow himself, first hazarded a guess at 7 K, which he later raised to 15 K, and on various different occasions he came up with even higher estimates, the largest being 50 K. When one remembers that the energy density of black-body radiation goes as the fourth power of the temperature one finds that a temperature of 50 K overestimates the energy density by as much as 11 000. Surely a prediction of this kind can hardly be considered precise. We will return to this point towards the end of this chapter.

The COBE and post-COBE measurements

The 1990 COBE spectrum is indeed remarkable in telling us how close the microwave background is to the black-body form. In fact it is argued that even in the laboratory one cannot generate such a precise black-body spectrum. How did the relic radiation get into such a precise black-body form? The big-bang scenario has the following explanation to offer.

As we saw in our discussion of primordial nucleosynthesis (Chapter 11), the early universe was very hot with particles of matter moving fast. In particular the light particles, mainly the electrons, were moving with speeds approaching the speed of light. These particles are very efficient in scattering radiation, that is the particles of light, the photons. And this scattering results in the radiation

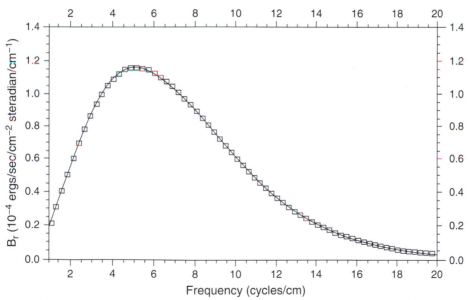

Fig. 12.6. The spectrum of the microwave background measured by the COBE satellite is shown above. The curve drawn through it is that of a black body that best fits it. The squares around the curve represent observational errors, which are quite small, thus showing that the fit is very good. The temperature of the black-body radiation is 2.725 K with a very tiny uncertainty of measurement of 0.002 K.

acquiring a black-body form. However, when the universe gets as old as, say, a few 100 000 years, its expansion slows down and the electrons also slow down. In fact they slow down enough to get attracted and trapped by *protons* which are heavier particles with opposite electric charge.

Electromagnetic theory tells us that oppositely charged particles attract each other. So the protons all along had the tendency to attract the electrons. However, while the electrons were very energetic and moving fast, the protons could not trap them and hold them. By the time the universe aged a few 100 000 years the electrons could no longer avoid becoming bound to the protons. And when they did, the proton–electron combination was none other than the familiar atom of hydrogen.

So when free electrons become trapped to form hydrogen atoms, they were no longer able to scatter radiation. This trapping stage is identified as the *epoch of last scattering*. After this epoch the radiation could travel freely through space without having to change its initial direction.

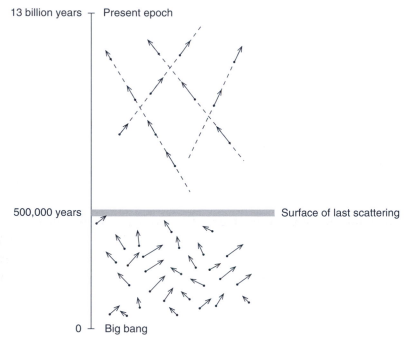

Fig. 12.7. This figure shows how far a photon gets without scattering in the universe. In the early times scatterings were very frequent and the photon could not coherently travel far. In later times scatterings were greatly reduced and the photon had a long free path. The changeover came relatively swiftly, thus identifying an epoch of 'last scattering' shown in the figure when the universe was 300 000 to 500 000 years old.

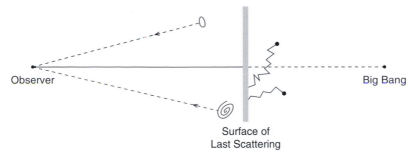

Fig. 12.8. The past light cone of a present observer is shown with the surface of last scattering. No event prior to this surface would be visible to the observer as photons coming from it are scattered and prevented from reaching the observer.

This was a major physical change in the state of the universe as far as astronomy is concerned. For, when an astronomer 'sees' a distant galaxy, he or she is able to do so because light from the galaxy reaches his or her telescope *undisturbed*. Had the particles of light emitted by the galaxy been scattered frequently, they could not have reached the astronomer's telescope carrying the information about the source galaxy. In other words, present day astronomy is flourishing because, as per the scenario painted above, observations of the distant parts of the universe mostly relate to events that occurred well past the epoch of last scattering. If, however, astronomers get more ambitious and their future instruments allow them to observe much further, they should, in principle, encounter a thick screen prior to the epoch of last scattering. In short they would be limited by a screen that prevents them from probing what the universe was like before the age of 100 000 years.

This is a major hurdle as far as direct observations are concerned. Basically it tells us that any probing of the early stages after the big bang is bound to be *indirect*, and hence more speculative. Nevertheless, this has not stopped the big-bang supporters from resorting to such probes, pushing their understanding well beyond the range accessible to astronomers. We will be devoting the next two chapters to these ambitious attempts. The detailed studies of the microwave background radiation play a key role in many of these exercises. In fact the 1990 success of COBE was followed by another discovery two years later, when the satellite discovered tiny ripples in the otherwise smooth and evenly distributed background, ripples in temperature of the order of a few parts in a million.

This latter COBE discovery brought immense relief to the big-bang theorists. For they had various theoretical scenarios to test to decide which one would work, *provided the radiation background showed at least some small ripples*. The COBE finding was taken as a great confirmation of the big-bang

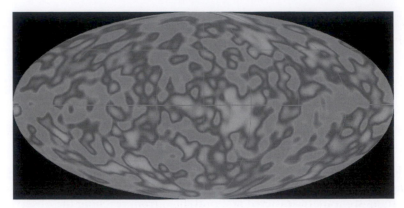

Fig. 12.9. The distribution of inhomogeneities of radiation on the sky first picked up by the COBE satellite in 1992. The dark and light patches indicate the ups and downs of temperature of the radiation from the average value. The departures were extremely small but were detected by the technology available to COBE. *Courtesy of NASA.*

scenario, with the team leader Smoot hailing the finding as 'seeing the face of God'.

All of the observations made since then have been designed to test various theoretical ideas all associated with the view that the radiation has arisen in an initial explosion. This is an unfortunate approach since none of the observers on the project ever consider the possibility that they may be wrong, as must be the case if there was no beginning. Thus they all analyze their data to try to fit their theoretical scenario, and when in doubt never consider the possibility that a very different approach might be worthy of study. We discuss this alternative in Chapter 15.

Since the big-bang model requires that the matter out of which the galaxies formed must have also originated in the initial explosion, in the theoretical scenario which has been developed it is assumed that there were primordial density fluctuations in the matter out of which the galaxies were formed by gravitational collapse, and also through mergers between different components.

As the expansion proceeded, the matter and radiation ceased to interact, and thus it is believed that the radiation detected today will contain indications that the early clumpiness of the matter put a gravitational imprint on the radiation field while it and the matter were strongly interacting.

The theoretical scenario also requires an effect called cosmic inflation in which the nascent universe underwent enormous growth about a billion-billion-billion-billionth of a second after the beginning of time. We will discuss inflation in the following chapter. It is claimed that this is now being supported by the data found by an orbiting observatory called WMAP (Wilkinson Microwave Anisotropy Probe; 'Wilkinson' in memory of David Wilkinson, who originally worked with Robert Dicke at Princeton to build a telescope to look for the microwave background).

WMAP is used to detect very small variations in the intensity of the microwave radiation with an efficiency far exceeding that of COBE. This scenario tells us that they are caused by the peaks and troughs of sound waves that echoed through the early universe and which themselves echo tiny irregularities in the distribution of matter that were frozen into the universal fabric by the sudden expansion, that is, inflation.

It is also claimed in this scenario that the universe is not what it seems to be to human senses. Normal matter which is seen in stars and galaxies is mostly made of atoms and molecules. The nuclei of atoms are known to contain protons and neutrons, which are collectively called baryons. Hence it is often said that normal matter is largely 'baryonic'. A series of arguments have been advanced that lead to the view that only a small fraction of all of the mass-energy in the universe – about 4%, is made up of normal matter. A second component, about 22% is dark matter. Some of this matter may be identical to the 4% which we cannot see because it is not emitting radiation, either because it is too old, made up of dead stars like white dwarfs, neutron stars or black holes, or is very young matter which has not condensed into stars and does not radiate. There is good reason to believe that dark baryonic matter is present. The well-established theory of stellar evolution shows us that this is plausible.

However, as we will describe in Chapter 14, in the popular big-bang scenario, another type of dark matter, is invoked. This is non-baryonic dark matter (NBDM) for which there is no observational evidence. However, without it the theorists do not know how to make galaxies. With it they are able, using numerical simulations, to begin to understand a complex universe.

But this is a classic example of modern cosmological thinking. Since in the big-bang scenario it is impossible to make galaxies without invoking the premise of non-baryonic matter, it is simply argued that it *must be there*. The possibility that the whole theory is wrong is never entertained.

The remaining component in the above balance sheet is called *dark energy*. The observational basis for believing that such a component exists, is that in the late 1990s it was shown that the expansion of the universe is not slowing down, but is accelerating. In all classical big-bang models with a zero cosmological constant the expansion slows down as the age increases. Only in the classical steady-state cosmology was it predicted that the universe should accelerate, essentially because mass-energy is being continuously created during the expansion.

The big-bang believers having found the acceleration simply ignored the steady-state prediction and immediately invoked a universe with a positive (non-zero) cosmological constant, and began to use the term dark energy without mentioning the alternative of creation. We will discuss all these flights of fancy in the next two chapters.

We, however, end this chapter with a tantalizing idea for the reader to mull over.

Could microwave background be relic starlight?

In 1955, fully ten years before the discovery of Penzias and Wilson, the three authors of the steady-state theory (*see* Chapter 9), Bondi, Gold and Hoyle had an internal argument. They noticed that stars like the Sun shine because they convert hydrogen to helium in a nuclear furnace. How much helium is so made by the Sun can therefore tell us how much radiation the Sun has sent out to date. So Bondi, Gold and Hoyle took stock of how much helium exists in the universe at present and used that information to calculate how much light was emitted by all the stars to date, assuming that *all observed helium was made in stars*. They came up with an energy density of such starlight, of 4.5×10^{-13} erg/cm^3. In 1958 one of us (GB) made a similar calculation which was published.

Gold conjectured that if all this relic starlight was frequently scattered, it would end up as black-body radiation of 2.75 K temperature! He suggested that the three of them write a paper predicting such radiation background in their steady-state universe. However, Bondi and Hoyle overruled him, saying that at the time they did not know of a mechanism by which the relic starlight could be made into black-body form. And so an important prediction which might have altered the future development of cosmology was missed.

We will take up this story again in Chapter 15.

13

The very early universe

Closer and closer to the big bang

The attempts in the latter half of the 1940s by George Gamow and his colleagues, described in Chapter 11, were only partially successful in explaining the origin of chemical elements. Of the 200 or 50 isotopes of atomic nuclei, only about half a dozen of them could be made in the hot furnace of the universe within a couple of minutes after the big bang. Even that could be achieved only provided rather finely tuned conditions relating density of baryons to temperature existed during that early epoch. These ideas of primordial nucleosynthesis were not very popular amongst the physicists who considered cosmology in the 1950s to be a highly speculative subject. This was one of the reasons why no attempts were made to actually look for the relic radiation predicted by Ralph Alpher and Robert Herman. The discovery by Penzias and Wilson in 1965 therefore came as a 'surprise' to the astronomy and physics communities.

As we also saw in Chapter 11, work on the alternative scenario of making chemical elements in stars achieved considerable success. The pioneering work of the Burbidges, Fowler and Hoyle, which we will refer to as B^{2}FH, showed that *almost all* of the isotopes can be made inside stars with predicted abundances agreeing with those observed. Besides, the typical star in this respect is a thermonuclear reactor which can be observed in detail with different masses, sizes and luminosities. To a physicist therefore a star is an experiment that can be repeated. Science sets great store by repeatability of experiments. Only if an experiment is performed several times under controlled conditions and its results observed to be consistent with the theoretical prediction, can the scientist be confident of the theory. By contrast, the primordial nucleosynthesis advocated by George Gamow happened *only once* and that too under finely tuned conditions and over a period not exceeding three minutes. Further, as

185

we saw in Chapter 12, the epoch of this event was so far back in time that it cannot be observed astronomically. That is why stellar nucleosynthesis was a scientifically more attractive solution to the problem of the origin of elements.

There were a handful of exceptions which posed a challenge to this approach, however. The light nuclei, like helium, deuterium, lithium, etc., were the nuclei whose abundance was difficult to explain. For example, if one went by the current stellar activity in the Galaxy and extrapolated it to the estimated lifetime of the Galaxy of, say, 10–13 billion years, the amount of helium produced by stars would be barely ten percent of the observed amount. By contrast if one went by the primordial route of George Gamow, one could get the right amount of helium by suitably adjusting the physical parameters of density and temperature. The question therefore was, whether stellar nucleosynthesis would eventually be able to explain the abundances of *all* of the nuclei, or was primordial nucleosynthesis *essential* to the process?

The finding of an isotropic radiation background at 7 cm by Penzias and Wilson in 1965 tilted the balance of judgement by most cosmologists in favour of the second option. For, it readily provided the explanation of the observed radiation as a relic from the primordial nucleosynthesis era. Belatedly the Alpher and Herman prediction was resurrected and it provided further support to this choice. During this and the following chapter, we will follow this line of reasoning which has indeed guided the course of cosmology in the post-1965 period.

As mentioned before, Gamow's primordial nucleosynthesis programme required us to study the big-bang universe from around one second to three minutes after the explosive origin of the universe. Taking that as the baseline

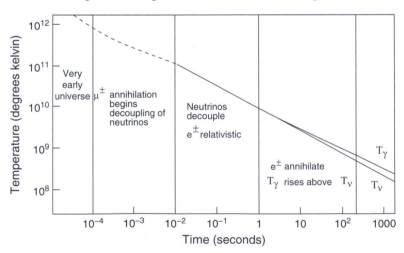

Fig. 13.1. The variation of temperature of the early universe shown at time elapsed since the big bang. The universe cools down very rapidly in early times because of its rapid expansion. Different physical processes take place at these early epochs as shown in this graph.

of investigations, cosmologists wished to go further back into time, getting closer and closer to the 'forbidden' singular epoch of big bang. The universe, of course got hotter and hotter along the way! The temperature at the end of the primordial nucleosynthesis era was estimated to be around a few hundred million degrees while the temperature at one second after the big bang was expected to be just over ten billion degrees. If one went even further into the past, the temperature got steadily higher.

We may talk loosely of the Gamow nucleosynthesis era as that of the 'early universe', and anything prior to that as belonging to the 'very early universe'. There is a fairly simple ready reckoner that tells us the temperature of the universe at any given time during this very early era. Start with ten billion degrees at one second. For every tenfold increase of the temperature, you have to go a hundredfold division of time. Thus at a hundredth of a second the temperature was a hundred billion degrees; at a millionth of a second after the big bang the temperature was around ten thousand million degrees.

What do these very large and very small figures mean? Unless one attaches physical interpretation to these numbers, they are meaningless. We will examine next how cosmologists collaborate with particle physicists in conducting such an exercise.

Astroparticle physics

The collaboration between cosmologists and particle physicists is motivated by the following concern. The temperature of a body or a region indicates the level of dynamic activity taking place there. When we heat water, the growing heat results in convection currents being set up with water globules moving up as they get heated. In a piece of metal the effect of heat is to increase the vibrations of its atoms and molecules. In short, the higher the temperature the greater the motion.

If we consider a volume containing gas which is being heated, the dynamical activity in it will grow. Mathematicians who study these motions find that there is no fixed speed with which gas particles move. There is a variation of speed from particle to particle, from the very small to the very large, and there is randomness in their directions of motion. These keep on changing as particles collide and scatter in different directions, but the average speed suitably computed can be related to the gas temperature. This calculation is possible because one is looking at an assembly of a large number of particles and statistical techniques can be applied with a high degree of confidence. The subject describing the collective motion of a large number of members of a population is called *statistical mechanics*.

It was statistical mechanics applied to a mixture of large numbers of electrons, positrons, photons and neutrinos as well as neutrons and protons, when the universe was one second to three minutes old, that enabled physicists to compute the formation of nuclei à la George Gamow. The temperature of the mixture at any stage during this era gave physicists information on the range of speeds of different particles of the mixture. There is a simple formula that tells us that the average energy of motion per particle is proportional to the temperature of the gas. For example, at the temperatures operating from one second to three minutes, the range of energies of particles corresponded to one where they would be subject to nuclear forces. The speeds were of such order that physicists were confident that they could apply their *known* laws of nuclear forces to study their interaction.

However, when one tries to go further back into time, to temperatures of the order of a billion billion degrees or even more, one is dealing with particles moving much faster than at primordial nucleosynthesis and one therefore needs appropriate physical laws to understand how particles moving so

Fig. 13.2. The particle accelerator at CERN, Geneva is in the form of an enormous ring as shown above through which particles are made to travel at faster and faster speeds. Credit: *CERN*.

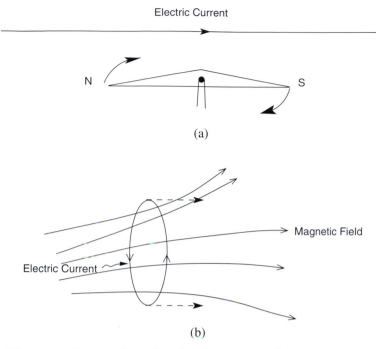

Fig. 13.3. These experiments show that the phenomena of electricity and magnetism are related. In (a) we have the effect of an electric current flowing in a wire on the magnetic compass needle. The needle turns as if moved by a magnetic force arising from the current. This effect was discovered by André Marie Ampère. In (b) we have a closed loop moving across a magnetic field. The loop has a transient electric current flowing through it as it moves. This effect was discovered by Michael Faraday.

rapidly will interact. The laws of nuclear physics do not help at this stage. One needs guidance from the emerging part of physics that deals with very high energy particles. Particle physicists who specialize in this field rely largely on the experiments conducted in particle accelerators.

A major achievement of twentieth century physics was to identify four fundamental forces in physics that appear to describe *practically all* observed properties of matter. These are (1) the force of gravity that was first identified as a law of nature by Isaac Newton in the seventeenth century, (2) the electromagnetic field which describes the forces between electric charges and currents, (3) the weak force that is behind the decay of such nuclear particles as the neutron and first became known as the phenomenon of *beta decay* and (4) the strong force which binds the atomic nucleus. A characteristic feature of these interactions is the energy of the participating particles. Another is the strength of their interaction. There is considerable difference between these basic interactions vis-à-vis these properties.

To begin with, gravity is the weakest force. To compare its strength with electromagnetic force, we compare the force of electrical attraction between the atomic particles, electrons and protons, with the force of their gravitational attraction. We find that the former is something like

10 000 000 000 000 000 000 000 000 000 000 000 000 000

times the latter! We may alternatively state that the latter force is a fraction

1/(10 000 000 000 000 000 000 000 000 000 000 000 000 000)

of the former.

Since we anticipate using very small fractions or very large numbers like the above in what follows, it is better to evolve a compact form of expressing them. The large number (with 40 zeros after 1) that we have above, can be written more compactly as 10^{40}.

Likewise, the fraction which is obtained by dividing 1 by this large number is written as 10^{-40}.

Using this notation we will write a million as 10^6 and a billion billion as 10^{18}. And a billionth part of unity will be written as 10^{-9}.

A holy grail for many particle physicists is the 'ultimate theory' which combines all aspects of these four separate ones. Progress in that direction can be seen in the following way. In the nineteenth century the forces of electricity and magnetism seemed inter-related. Magnetic effects were produced by electric currents and changing magnetic fields lead to electric currents in circuits. Eventually, in the 1860s, James Clerk Maxwell came up with a theory that properly unified these two forces. The successes of this theory led to false optimism that the end of physics was near. However, further studies led to two major revolutions in physics: the quantum theory and the theory of relativity. New forces of nature were discovered. A century later, in 1971, Abdus Salam and Steven Weinberg showed that the electromagnetic theory of Maxwell can be unified with the weak interaction to produce the 'electro-weak' theory. This conjured up the vision of a 'grand unified theory (GUT)' in which the electro-weak theory may be combined with the strong interaction.

Where does gravity fit into the scheme? In fact the idea of unification of basic interactions did not originate with particle physicists of today, but owes its conceptual origin to Albert Einstein. It was back in the third decade of the last century that Einstein proposed that the electromagnetic theory is combined with gravity in what he called a *unified field theory*. Although he made several attempts followed by others, the project did not materialize. Even today, gravity seems more difficult to bring under one fully unified framework.

Nevertheless towards the end of the 1970s, the success of the Salam–Weinberg approach raised considerable hopes of a grand unified theory of the other three interactions. The optimism was typical of the view expressed by Stephen

Hawking in his inaugural lecture for the Lucasian Professorship that 'the end of physics is round the corner'. Part of this optimism was fuelled by the emerging possibility of a collaboration between particle theorists and big-bang cosmologists.

A major problem facing the GUT programme was the energy of participating particles. This energy is usually measured in units of *Giga Electron Volt* (shortened to GeV). Leaving aside the technical definition of this unit, we may try to understand its magnitude with the famous Einstein equation $E = Mc^2$...the energy obtainable at the expense of mass. If we manage to destroy a proton, the energy we would get in return would be very close to 1 GeV. Modern accelerators which produce particles like the proton with very high energies, reach up to the level of 1000 GeV.

In the case of the unification of electromagnetic and weak interactions, the consequences of the unified theory needed to be tested at energies of the order of 100 GeV. This was possible at the accelerator at CERN (Conseil Européenne pour la Recherche Nucléaire) in Geneva. The energy at which definitive evidence for the operation of the electro-weak theory was obtained could thus be comfortably generated by this accelerator.

The problem of testing similar predictions of a GUT is put in proper perspective by the fact that the energies of particles that need to be produced for this purpose are of the order of 10^{16} GeV, about ten thousand billion times the highest energy the best man-made accelerator can produce. The gap is wider than that separating the distance to the Moon from a foot rule. Since in science a theory gains credibility only by passing some crucial experimental tests, any theory like GUT would necessarily fail this criterion. Unless...

Unless astronomical evidence can be brought to bear on the issue. Astronomy has provided evidence for the operation of laws of physics on a scale grander than any possible through laboratory experiments. The motions of planets, satellites and comets confirm the validity of the law of gravitation, for masses far greater than any possible in a laboratory experiment. The operation of a controlled nuclear fusion reactor, so far not demonstrated in a terrestrial laboratory, is well demonstrated inside stars. Can astronomers provide sites of ultra high energy particles where the GUT can be seen to operate? Unfortunately not, if one goes by the conventional astrophysics that talks about stars and galaxies. The energy of particles produced in an exploding star (called a *supernova*) reaches a level of 1 GeV or so. Quasars, described in Chapter 10 may eject particles of energies 1000 times or a little higher, but by no means close to the energy needed to verify GUTs.

However, there is one exceptional situation where high enough energies can be reached. In the big-bang cosmology, as we saw earlier, we can reach energies as high as we want, provided we go sufficiently close to the big bang. By our rule of thumb, if we were willing to venture within a time span of 10^{-37} second of the big bang, we would have encountered particles with high enough energies to have participated in a GUT-type interaction.

Lest the reader thinks the time span 10^{-37} second too short, there is an even shorter time span of 10^{-43} second, that is a millionth of the GUT time span, when quantum gravity was believed to dominate. This, the earliest period so far, shows that the rule governing the behaviour of gravitational force was dictated not by Newton or Einstein, but by quantum theory. As yet this is an unknown frontier which no physicist has successfully mastered.

Notice that even the GUT time span is extremely short – in words it is a tenth of a billion-billion-billion-billionth part of a second. As of now, the shortest time span that can be operationally measured in a lab is around 10^{-13} second, and that is done by a caesium clock. Astronomers using the extremely regular pulsating stars called *pulsars* can achieve ten times better accuracy in time keeping. However, these capabilities fall far short of measuring the extremely short time span after the big bang when GUTs are supposed

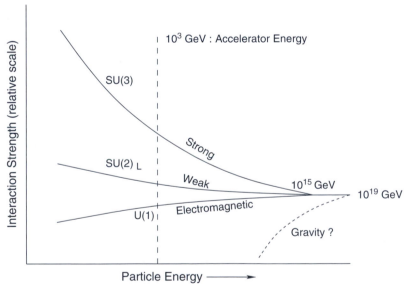

Fig. 13.4. This chart shows how powerful the different basic interactions at various particle energies are. At lower energies, the strong interaction is the most powerful. At energies approaching the value 10^{16} GeV, particles experience all three basic interactions with comparable strengths and this is the energy for 'grand unification'. The best modern accelerators can produce particles of energy not exceeding 1000 GeV. So the gap between what can be confirmed by experiments and what theorists speculate is very large, represented by the factor 10^{13}.

to come into play. Thus we have only a theoretical concept of time to fall back upon.

At this epoch, which we may term the 'GUT epoch', the temperature of the universe is said to be around 10^{27} degrees (that is, around a billion billion billion degrees). Yet even the normal usage of statistical mechanics, which relates the concept of temperature to average energy, breaks down. It turns out that the characteristic size of the universe is so small that there are not enough particles in it to justify the use of statistical mechanics.

If we turn a blind eye to these conceptual problems, we have a situation wherein we *can* apply the concepts of ultra high energy particle physics and explore their consequences. The particle physicists are therefore attracted towards the big-bang scenario, since it supplies the only opportunity for the application of their ideas on grand unification. For the big-bang cosmologists also it is worthwhile collaborating with the particle physicists, since they are the only source of physical ideas that can be used to know how the universe behaved so early in its lifetime.

The subject that has grown out of this alliance between particle physicists and cosmologists is popularly known as *astroparticle physics*. The impression is created that with a GUT around the corner and the details of the early universe well understood, the 'end of physics' is not very far. We will explore this belief and some of its consequences but with the following caution.

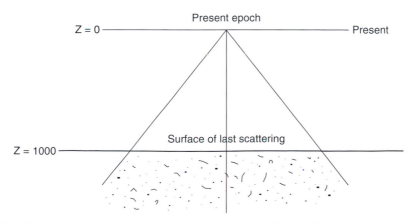

Fig. 13.5. This figure reemphasizes the point made earlier in Chapter 12 that frequent scatterings of the photon prevent information from the very early epochs in the universe from reaching the observer. The 'surface of last scattering' thus stands as a screen for the present observer, a screen that blocks the view of what happened in the universe prior to that epoch. Objects at this surface would show redshifts of the order of 1000.

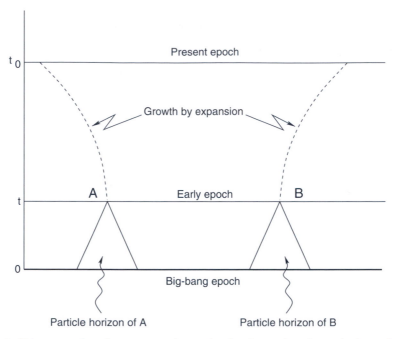

Fig. 13.6. Diagram showing space along the horizontal axis and time along the vertical. Close to the big-bang epoch, consider two observers A and B. Their past light cones are shown. These represent the particle horizons of A and B. Note that they do not intersect. This means that they are causally disconnected, that is, their physical evolution is independent of each other. If we require the universe to evolve into a homogeneous isotropic form, its component points need to be causally connected. This requirement means that these points have to be very close to one another. Calculations, as discussed in the text, then show that with this restriction the universe at present cannot be homogeneous on a scale larger than a few centimetres. This is known as the *horizon problem*.

The particle physics that is used for these studies is *non-proven*. In fact, the object of the whole exercise on the part of particle physicists is to find a scenario where their theories can apply and the very early universe provides the one and only such scenario.

The cosmological part of these studies is also speculative, since as we saw in Chapter 12, the cosmic microwave background erects an opaque screen in time prior to which nothing can be observed astronomically. Cosmologists claiming to know what happened before the epoch of last scattering are like spectators at a play guessing what is happening on stage behind the curtain which is down.

Astroparticle physics therefore rests on two highly speculative pillars and any conclusions drawn from it are subject to this caution. Lest the reader is carried away by the apparent successes of this approach, we will have occasion to remind him of this situation from time to time.

Three conceptual problems

When exploring the very early universe, based on extrapolations of our present understanding, one encounters several conceptual problems apart from the very basic issue of measuring time intervals as small as 10^{-37} second. These are called the *horizon problem, the flatness problem* and *the monopole problem.* Let us see what these problems are.

Imagine, that we set the clock ticking with the big bang and think of an observer who makes studies of the universe when it was only one year old. As he looks further and further away, he also looks back in time. An object half a light year away would be seen by him as it was half a year ago. By the same token, we find that the *furthest* he can see is only one light year away. . .because then he is looking at the state of the universe one year ago, that is, at the primordial epoch of the big bang. He *cannot see* anything beyond one light year away since there was no universe to look at then.

We say that such an observer is limited in his viewing of the universe by a horizon sphere of radius one light year. The horizon is sometimes called the *Particle Horizon.* We can go by the analogy of the spherical Earth. Someone on the top of a tower looking far away is limited in how far they can see, by the fact that the Earth is curved like a sphere. (See the figure illustrating this.)

Apart from mere viewing limitation, the particle horizon has a deeper significance. It also limits how far any cause and effect chain will go. A causal

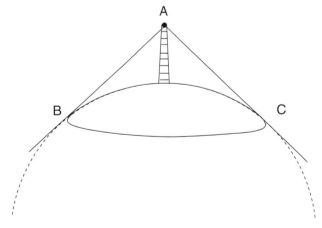

Fig. 13.7. The observer on the lighthouse tower is limited in his view of the surface of the ocean, because of the curvature of the Earth. Only the region BC, above the dotted line will be visible to the observer.

effect will at most propagate with the speed of light; for special theory of relativity tells us that no physical influence shall travel with speed greater than that of light. So we can imagine the universe at the age of one year, divided into separate little spheres each of radius one light year, so that there is no communication between the centres of two spheres.

At this stage we consider the analogy of old human habitations, say, several thousand years ago. As there was no fast communication on the Earth at the time, these habitations were more or less isolated and grew in their own different ways. In today's modern era of fast communication isolation is no more and different parts of the world learn of each other's way of life and there is a tendency towards a much greater degree of uniformity of lifestyle.

In the same way we would have expected the universe at an age of one year to have a typical size of one light year over which it would be uniform. We also see that as the universe grows older there is a tendency towards much greater uniformity as the particle horizon grows in size.

However, in the case of the big-bang universe, the era when the uniformity was fixed came much earlier. It should have come at the GUT era if not earlier. Because, at that stage all interactions were working in a unified way and determining the state of matter on which future development would be based. And that era was around the age of 10^{-37} second. We should be checking up on the sizes of the horizon spheres at this epoch. Multiplying by the speed of light to this time scale gives us a distance scale of 3×10^{-29} metre; that is a distance you would get if you divided the length 3 metres by the large number a hundred billion billion billion. When the universe was trying to get itself into shape, its typical size of uniformity was no larger than this.

That was at that early stage. Since then it has expanded considerably. Theoretical calculation tells us that the universe has expanded since then by a factor 10^{27}, so that our sphere of uniformity has also grown in size by the same factor. If we do a little arithmetic multiplying the very small number 3×10^{-29} metre by the very large number 10^{27}, we get a length of merely 3 centimetres! This means that we should not expect the universe to be uniform and homogeneous on a scale greater than 3 centimetres. As we saw in Chapter 12, the microwave background is so homogeneous that the overall size of uniformity is as high as 10^{26} metres. Thus instead of getting a very homogeneous universe on such a large scale, our theory predicts a very patchy cosmos. Clearly something is wrong in our theory, wrong by a large margin.

This is the *horizon problem* which essentially tells us that the physical features of the universe could not have been determined at the very early GUT epoch suggested by astroparticle physics. This fact had been pointed out by theoreticians in the late 1960s after the discovery of the cosmic microwave background, but had been ignored by most since they did not have an answer for it. Certainly the big-bang models of Friedmann and Lemaître seemed to fail miserably and by a large margin.

The second problem is commonly known as the *flatness* problem. To understand it we have to go back to Chapter 7 where we found that the nature of the geometry of the universe is closely related to the density of the universe. If the density happens to be equal to a critical value, the geometry of space is of a *flat* Euclidean nature. If the density of matter is greater than the critical value, the universe is *closed*, while if the density is less than the critical value, the universe is *open*. Present observations show that the density of matter in the universe is comparable to the critical density, maybe ranging from a few percent of it to a few times it. We will examine the actual evidence in the next chapter.

Now, it is commonly assumed that the nature of the geometry of the universe was fixed at an early era like the GUT era. The difficulty that such an assumption leads us to is illustrated by the following example.

Imagine a person is attempting to fire arrows at the target using a bow for shooting. It is clear that if he stands very close to the target, the chance of his hitting the bull's eye is large. Even if he makes errors, the arrow lands fairly close to the central circle. As he moves away from the target, the frequency of his errors will increase and the arrow may land away from the bull's eye quite often and also the distance where it would land will also increase. In fact at a distance, of say 5 metres or so, it will take a very accomplished player to shoot the arrow within a small distance of the target.

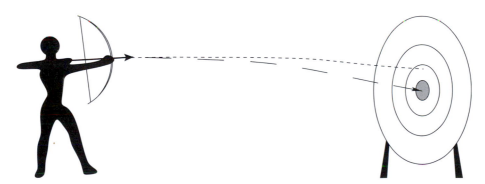

Fig. 13.8. For an archer, the problem of hitting the bull's eye gets more and more difficult as his distance from the target increases. He has to aim very precisely at the bull's eye, if he is to get his shot right: the margin of error allowed becomes narrower and narrower. Likewise, the density of matter in the early universe had to be very finely tuned in order that its present value is within the observed error-bars. This tuning gets finer and finer if the fixation of density took place closer and closer to the GUT epoch.

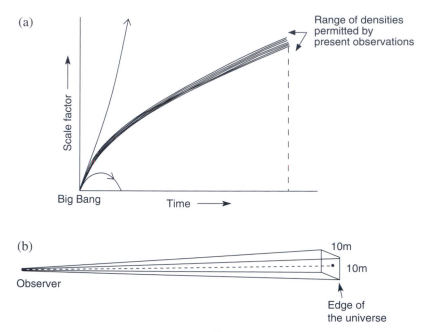

Fig. 13.9. The fine tuning of one part in 10^{50} (expressed in an exaggerated form by the grey strip in the upper diagram) can be understood by the example of the arrow shooter. It is comparable to the chance of hitting a 10 metre × 10 metre target located at the edge of the presently observable universe, if one picked out the direction at random.

Likewise, suppose physical conditions at the GUT epoch conspire to decide the density of matter in the universe. The target is to attain closure density. If there is a small error and the density is *less* than the closure density the universe would have an open geometry and it would expand fast...so fast that at the present epoch there will be no other galaxy left in our neighbourhood to observe. If the density achieved turned out to be greater than the closure density, even greater catastrophe would be in store. The universe would slow down and collapse into a big crunch and we would not find it expanding at the present epoch! Only if the margin of error were very small, would we find the universe with present density within, say a factor of 1/10 to 10 of the present closure density. How small should the margin of error be? The calculations show that the margin of error permitted is smaller than one part in 10^{50}. In short, the density of matter has to be very finely tuned close to the critical density at the GUT epoch if we wish to find the universe to be in the state observed today.

Such fine tuning is hard to accept in a physical theory. Why, within a large range of possibilities offered to the universe at the GUT epoch, did it choose one that lay in a very finely tuned range around the critical density?

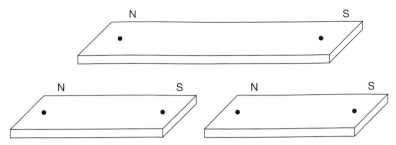

Fig. 13.10. If a bar magnet is cut in half as shown above, it breaks into two bar magnets, each with the two opposite poles. In short, one cannot hope to isolate the poles from each other by the cutting operation.

The horizon and the flatness problems were known for some time in the 1960s and 1970s, but the third, the *monopole problem* surfaced only when the grand unified theories began to be talked about. The problem owes its name to the concept of magnetic monopoles. We know that a bar magnet has two poles, the north and the south. They appear to be together in a pair. Yet if we wish to isolate them, say, by cutting the magnet in two parts, we do not succeed! The two smaller bars end up having two poles each. The combination of pairs is called a magnetic dipole. Maxwell's equations, which unified electricity with magnetism, tell us that nature does not permit singleton magnetic poles to exist. By contrast there are singleton electric charges, both positive and negative, which exist in abundance. For example, an electron is a carrier of negative unit charge while a proton carries a positive unit charge.

Going beyond Maxwell's theory to a typical grand unified theory, the conclusion that particle physicists arrived at was rather disturbing. Given the general symmetries that GUTs followed, the conclusion emerged that magnetic monopoles must exist in isolation. Not only that, a typical monopole would have a mass of the order of 10^{16} times the mass of a proton, and a monopole once created cannot be destroyed. *This meant that all those primordial monopoles created around the GUT era must be around today.* This conclusion would have been tolerable if the numbers of those relic monopoles were very small. . .so small that their presence or otherwise would have made no difference. However, calculations showed that this was far from the case. The density of matter in the form of relic monopoles today would have been at least a million billion times the critical density of matter. For monopoles to have remained without detection of their magnetic attraction and also with their huge mass density is impossible.

This problem seemed to be insurmountable and drove the cosmologists to the prima facie conclusion that collaboration with particle physicists implied

GUTs, and GUTs implied magnetic monopoles and these were simply not acceptable on present day cosmological evidence.

Enter 'inflation'

This situation existed till the end of the 1970s when an apparent resolution of these problems was offered by Demosthenes Kazanas, Katsuhiko Sato and Alan Guth, all independently arriving at the idea known today as an 'inflationary universe'. The terminology is that of Guth. Ironically the original model proposed by Guth turned out to be faulty, but the idea of inflation survived, largely because it was marketed superbly. We will first describe the basic scientific idea and then make comments on the related sociological issues.

Recall that the difficulties we mentioned above with the standard big-bang scenario seem to be related to the circumstance that the universe was too small in size at the time it was 'hot' enough to carry particles energetic enough to participate in the grand unified interaction. Could some way be found to 'blow' it up to a very large size? Recall that the principal but mysterious source of such blowing up, the 'big bang', had already happened and there was no way one could invoke it again. So rather than take recourse to mystery, this time an attempt was made to appeal to physics. And the GUT era held out hope of generating extra energy to increase the size of the universe rapidly in the following way.

First we look at the epoch when the typical particle energy was equal to the GUT energy of around 10^{16} GeV. This energy is expected to be one when all three participating interactions (strong, weak and electromagnetic) are of comparable strength. In an expanding universe, the typical particle energy will fall with time. So, the basic interaction controlling the behaviour of matter would change over from GUT to the separate constituent interactions, strong and electro-weak, as the typical particle energy fell through this value of 10^{16} GeV. Such a changeover in the nature of the physical state of matter is called 'phase transition'.

We encounter phase transitions in several natural phenomena. The commonly observed changeover from water to steam as one heats water, or vice versa as one cools it, is an example of phase transition. Some phase transitions require energy to bring them about, while some release energy. It is the specific physical process that decides which way the effect lies. For example, water needs to be supplied with heat energy before it can be converted to steam while it releases energy on freezing to ice.

Normally, when we heat water its temperature rises. However, the rise goes on until it starts boiling, when the temperature is 100 degrees Celsius or 212 degrees Fahrenheit. If we continue to supply heat the temperature does not rise; instead water is converted to steam. So steam possesses that extra heat energy which makes the difference from the liquid state (water) to the gaseous state (steam). Since the temperature remains the same, one can say that this energy is some kind of hidden energy...energy that is utilized to convert the fluid state of matter into the gaseous state.

The energy difference between water and steam is called the *latent heat*. To convert one gram of water into one gram of steam the latent heat needed is 536 calories, about the energy one consumes through a moderate steak and potato meal.

Suppose we reverse the process and start cooling steam. As its temperature hits the 100-degree mark, and we continue cooling it further, it will first change over to water, without any lowering of temperature. At this stage, the cooling process releases the latent heat of steam. (Even the refrigerator releases heat at the back as it cools whatever is stored inside it.)

There are situations when the cooling of steam can be carried out in such a way that its temperature falls below the 100-degree mark, *without it changing to water*. Such specially cooled steam is called 'supercooled' steam.

This state of steam is unstable for the following reason. Nature, as a rule, operates its laws in such a way that any physical system tends to take up the lowest state of energy possible. That is the state that is stable and permanent. A state with greater energy will sooner or later change over to the lower energy stable state. The example of a pencil standing on end (see figure) illustrates the idea. The 'standing' pencil has higher energy but is unstable: it falls and lies on the floor in a state of lower energy, a state which is stable.

Returning to supercooled steam, it has two possible states: its existing state in which it still retains its latent heat energy, and the state of water in which the latent heat is absent. By our rule just stated, the steam state is unstable and like the standing pencil, it will collapse into the liquid state of water. By 'collapse' we mean it will begin to condense as hot droplets of water. And in such a collapse, the extra energy is released.

Physicists believe that such an energy release may have occurred when matter in the universe cooled through the GUT epoch. Rather, matter may have 'supercooled' in a state that existed before the GUT phase transition took place. And, after a brief span of time, it realized that the state it was in was not the stable state of lowest energy and so it changed over to that state spontaneously...just as supercooled steam condenses into droplets of water.

The changeover from the higher to lower energy state releases the latent energy. And this is the energy that was supposedly used to power the universe so it expanded rapidly. Since the original post-big-bang universe was expanding anyway, this rapid expansion phase is called *inflation*.

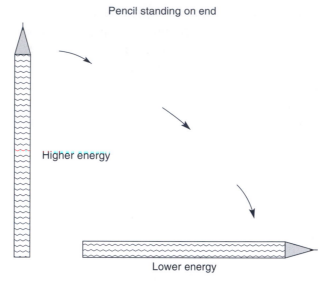

Fig. 13.11. A pencil standing on its flat end is unstable and will fall to a horizontal position as shown if disturbed slightly. The force of gravity induces the fall. The gravitational energy of the pencil in the standing position is greater than in the horizontal position. This is an example of the natural tendency of a system to attain the state of lowest energy.

The difference between ordinary expansion and inflation can be understood by looking at the analogy of interest rates. Suppose a person places a sum of $1000 in a bank deposit which offers a simple interest of 5% annually. Suppose he simply forgets that he invested this money. After one hundred years, the bank discovers it and informs his successors that they may claim the deposit. By then it would have grown to $6000, being the sum of the capital $1000 and interest thereon of $5000.

But suppose, the money were deposited under the scheme of *compound* interest. This means that every year the interest would be added to the capital and the sum would form the new capital. That means, after one year, the capital would grow to $1050, since the interest of $50 was added to the original deposit of $1000. Then the following year the interest of $52.50 would be added to the new capital making it up to $1102.50. The process continues this way. Where would it take the deposit to at the end of a hundred years? A pocket calculator will tell you that the answer is the considerable sum of $131 501. The sum invested for 1000 years would grow to more than 1500 billion billion dollars. (Under simple interest it would have grown to only $51 000.)

The difference between simple and compound interest illustrates the difference between normal expansion and inflation of the universe. The latter grows the size of the universe far beyond what the former could have done, even if inflation lasted for a limited time. As it would! For once the energy released through transition after supercooling were exhausted, the inflation would stop and the universe would revert to its original slower rate of expansion. The figure illustrates this scenario.

This rapid expansion, and its end, will not occur all over the universe simultaneously. Rather, we would have a situation in which a small part of the universe would grow as a bubble arising inside a rapidly inflating universe, depending on the region where the extra energy was expended. One may then have the universe as made of several such bubbles occurring at different space and time with ourselves being inside a typical one of these. In many respects this idea may seem similar to the Hubble bubble of Hoyle and Narlikar in the context of their ideas on creation of matter. Indeed it is now recognized to be so.

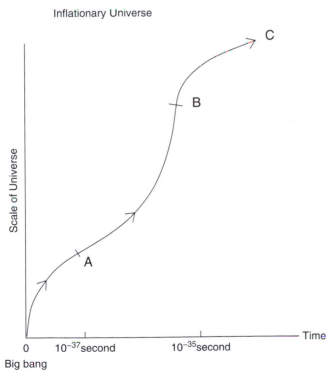

Fig. 13.12. The inflationary universe is characterized by a rapid expansion for a limited time as shown above. The scale factor increases rapidly between stages A and B. Prior to A and after B the universe is driven by a slower rate of expansion characteristic of a Friedmann model. The inflation is initiated at A by the transition in the state of matter and is over by the final instant B.

Is inflation really necessary?

The original idea of inflation as proposed by Alan Guth did not succeed in doing what it set out to do, although the reasons for this are too technical to go into. Nevertheless, the concept of an inflationary phase lasting something like 10^{-37} to 10^{-35} second was found attractive by many cosmologists. So new models of inflation arose. There are the 'new inflation', 'chaotic inflation', 'Kaluza–Klein inflation', 'eternal inflation' and several other ideas around. The eternal inflation is very similar to the much-maligned steady-state theory together with the Hubble bubble. However, the questions that a hard core scientist should be asking are: Is there any reason for big-bang cosmology to have such a phase? If admitted as necessary, what exactly is the physical mechanism generating it? If the physical mechanism is recognized and accepted, does it fit into Einstein's general relativistic framework? And, last but not least, what is the present evidence that unequivocally tells us that inflation did indeed occur?

We begin our answers to these questions with the first. We have already mentioned that the horizon problem, flatness problem and the monopole problem presented difficulties to the very early universe. Inflation justifies its existence if it can show that it solves these problems. Indeed it does. The horizon problem was posing severe restrictions to the sizes of homogeneous regions within the universe. By rapidly inflating the universe by a linear factor of around 10^{50} or so, these homogeneous regions will be made enormously large (perhaps larger than really required!).

The same trick solves the flatness problem. Imagine the analogy of a spherical ball. If the radius of the ball is, say, no more than one metre, then we can say that a circular disc of around one centimetre radius on the sphere would appear nearly flat. . .but a disc of radius twenty centimetres will not. However, if we blow up the sphere to a radius of one kilometre, then a disc of twenty centimetre radius will also appear flat on it. So one can say that because of extensive inflation of size, no special fine tuning of matter density is needed at the GUT-inflation era in order to arrive at the present 'nearly flat' geometry of the universe.

Inflation resolves the monopole problem by 'inflating them away'! That is, when the volume of any region containing monopoles is increased by a large factor, the number density of monopoles falls. The density falls to such a low value that we would hardly expect to find a single monopole in the observable universe.

The monopole problem was brought about by the grand unified theories. The problem was non-existent before these theories came about. However, having created the problem, its resolution was offered through inflation. The reader may notice a similarity between this example and the counter-Earth hypothesis of the Pythagoreans mentioned in Chapter 2.

These initial successes were responsible for many theoretical cosmologists being attracted towards inflation. However, a second look would show that clever marketing strategy has been at work: make the consumer feel that the product you wish to sell is essential for him by pointing out his sorry plight without it. None of these three problems were around in a serious way before forays closer and closer to the big bang were attempted. Earlier cosmologists had felt wary of application of physics very close to the big bang and so were not faced with these problems. All three surfaced with astroparticle physics and so the feeling has grown that once one opts to go closer and closer to the big bang, these problems will be bound to arise and their only remedy is inflation.

So we conclude that inflation is not essential to cosmology: it can be avoided if one did not have to go close enough to the big-bang epoch to encounter its perceived need. Turning to the second question, do we have a definite physical theory for it? Surprisingly, the idea of inflation is uncritically accepted by its supporters without there being a definitive physical theory for it. This is because, the GUTs or other ideas in very high energy physics have not yet zeroed down to any single formulation and hence several 'toy models' are in literature but no 'fully grown up' model accepted by all physicists. Therefore an objective assessment is that we do not yet have a theory of inflation that is on as firm a footing as, say, the electro-weak theory. This is the reason why there are many brands of inflation in the marketplace.

Coming to the third question, even if we assume that one of the toy models is the real one, there is as yet no rigorous solution to describe how the inflationary phase begins and ends by tying the loose ends of how the changeover occurs. In general relativity, great importance is attached to exact solutions since only they can provide credence to the physical description of reality. For, an exact solution tells us how the spacetime geometry is behaving both in time and space. Only when such a solution is known can one talk of any physics going on in the spacetime. Where such an exact solution is known to exist but is believed to be approximately applicable to the real life situation, the approximations can be used with confidence, as is done when applying to the description of the motions of planets in the Solar System. Unfortunately right from the early exercise of Guth, there has been no attempt on the part of general relativists to demonstrate that an inflationary solution with appropriate boundary conditions exists.

Coming to the final question, what is the true stamp of inflation that can be looked for in the present day evidence to convince ourselves that it did occur transiently very early on in the history of the universe? There inflation makes one very clear cut prediction. So close does it drive the cosmological model to the flat Friedmann model that it leads to the conclusion that the present day density of matter should be equal to the critical density. This critical density of matter as evaluated today is very close to 10^{-26} kilogram per cubic metre. Recall that at the end of Chapter 7, we used the symbol Ω to express the ratio of the actual to the critical density. Thus, what inflation requires is that this quantity Ω should equal 1.

We will examine this evidence in the next chapter, since in the last analysis, the proof of the scientific pudding lies in its test by experiments.

But we must *remind the reader* again that practically none of the ideas in this chapter are directly testable. Even worse the laws of physics which we always use must have been created in the first 10^{-43} second and have never changed in 10^{10} years.

14

Dark matter and dark energy

The emperor's new clothes

The Danish writer from the nineteenth century, Hans Christian Andersen wrote several fairy stories and folk tales, but none of them came closer to describing the current situation in cosmology than the well-known *Emperor's New Clothes*. We begin by recounting the story in brief.

An emperor was fond of trying new dresses and spent a fortune on various fashion designs. One day a couple of dressmakers from a far away land came to his court promising the emperor clothes made of such fine silk that only the virtuous and the righteous could see them. The emperor was pleased by this offer and accorded them liberal funds and facilities to make a royal dress. Taking considerable time over the process the tailors returned carrying their handiwork.

The king sent an emissary to examine the dress. When the packet was opened, the minister could see nothing in it. However, recalling the makers' admonition that only the righteous and virtuous could see them he felt that if he admitted to seeing nothing, he would be treated as a sinner and dismissed from his job. So he reported to the emperor praising the dress in glowing terms. Eventually the king decided to wear the new clothes himself and parade in them through the main street of his capital.

He too could see no dress; but the tailors went through elaborate motions of placing it on his body, commenting on how well it looked on his majesty, and he too felt that admitting seeing nothing would lead to his forsaking his kingdom as not being virtuous and righteous. So he got ready to join the procession followed by his courtiers all of whom were all praise for the new suit, since none wanted to be fired from his job.

As the procession went through the town people gathered on the street to applaud. Although they saw their emperor naked, they dared not say so for fear of being branded sinners. It was left for a simple child, who had no personal stake in the matter, to come out with the fact: 'Why, the emperor is not wearing anything!' That was when everybody realized that the emperor and his court had been taken for a ride!

With this introduction we return to modern cosmology, to perhaps its most important issue as to just how much matter and energy are present in our universe. And, how much of it we can see and how much we *cannot* see.

Evidence for dark matter

What is the meaning of 'dark' or 'unseen' matter? In the old days there was the adage: 'Seeing is believing'. This implied that only the evidence that you can see with your eyes can be trusted. The science of astronomy evolved through the process of 'observing' with the naked eye, and later with telescopes. Even when using the telescope for the first time, several people were uncomfortable with the findings made with its help, as they showed many more objects than were visible to the naked eye. Recall in Chapter 2 the doubts expressed about the reality of craters on the Moon, the spots on the Sun, and Jupiter's satellites when Galileo first demonstrated the efficacy of the telescope.

Galileo's telescope, and others that followed his pioneering instruments monopolized viewing to the form of light that our eyes are sensitive to. By the end of the nineteenth century, physicists were aware that electromagnetic radiation can come in other forms too with wavelengths vastly different from those which give visible light. The twentieth century gradually brought those other forms of light to the service of astronomy and 'seeing' now means using any of the different forms of light for observing.

So what do we mean by dark or unseen matter? It means the matter that cannot be seen but whose existence can nevertheless be inferred by indirect observations. The historical example of the discovery of the planet Neptune, described in Chapter 3 shows that the existence of Neptune was inferred by noticing its perturbing effect on the motion of the planet Uranus. Thus the existence of the new planet *could be deduced* even before it was seen in the conventional way. And the interaction that played a crucial role in the episode was gravitational interaction.

In modern times, gravitational interaction plays a similar role in revealing the existence of matter that could not otherwise be seen by using any form of light. It is this type of matter, and it may be in several different forms, that is labelled *dark matter*. How is it detected?

An analogy from the field of economics is worth recalling in this context. Think of a country which has two economic systems in force. The first is the

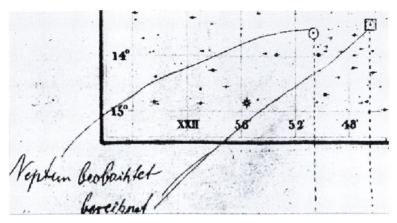

Fig. 14.1. The original Bremiker star-map from which Galle found Neptune. The position predicted by Le Verrier is marked by a square and Galle found an object, marked by a circle, missing from the chart. The source of this original Bremiker star-map from which Galle found Neptune is not traceable.

official one based on declared incomes and expenses; one which is on the records of the Internal Revenue Department. The second, parallel economy is run by the so-called black money, based on incomes and expenditures not reported to the taxman.

Now, even though the black money is not declared or recorded, experts can form a shrewd judgement of its extent. This is estimated by its visible impact on the country's economy. The construction activity, election campaign expenses, massive entertainment projects, etc. are the dynamical effects of black money...the economic activity generated by it. It is these that give the clue as to the amount of black money in circulation.

Dark matter in astronomy is like black money in economics. Although not directly observed, its gravitational influence on the visible matter in its neighbourhood can give astronomers good estimates of its total amount. Perhaps the most dramatic example of this type is the *black hole*. A black hole is a highly compact object whose gravitational pull is so strong that not even light can leave its surface. A very massive star may shrink under its own gravity and become a black hole when its surface gravity has grown powerful enough to pull back its own radiation. A black hole can therefore never be seen. Yet its gravitational influence will help to reveal its presence in space. For example consider a star having a planet going around it. If the star shrinks and shrinks and becomes a black hole, it will be invisible. Yet the planet will continue to feel its gravitational attraction and will keep orbiting around it. So if we see a planet going around and round but no star that is visibly controlling its

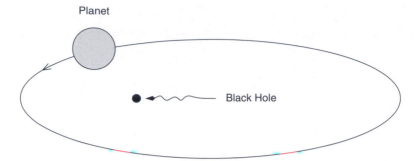

Fig. 14.2. The planet P in the figure will be seen travelling in an elliptical trajectory, but the star around which it is orbiting will not be seen if it is a black hole. However, from the law of gravitation and Kepler's laws, discussed in Chapter 2, the presence of the star could be deduced as being at one of the two foci of the ellipse.

movement, then we conclude that the planet is going around a black hole. By observing details of the planetary motion, theoreticians can tell where the black hole is located and what is its mass.

We now come to cosmological evidence for dark matter, mainly from two different types of systems.

Rotations of spiral galaxies

In Chapter 5 we described galaxies of which our Milky Way is one. It belongs to the class of *spiral* galaxies. We see one example in Figure 14.3. As the name implies, a spiral galaxy has two or more arms winding outwards like the spring of a classical wind-up clock. The arms are the regions where stars are concentrated. The gaps between arms are relatively less populated with stars, although they may carry gas and dust. The picture shown also indicates that there is no sharp boundary to the galaxy…it merges into darkness as one goes further and further from the more populated central region.

Astronomers believe (and justifiably so!) that the darkness engulfing the galaxy in the outward parts is indicative of its gradual but definitive approach towards a boundary. Thus they assume that beyond some specified perimeter, there is no mass belonging to the galaxy. Certainly there are no shining stars, nor are there any indications of gas or absorbing dust either beyond the assumed boundary. With the advent of radio astronomy, however, it was discovered that there are small or large clouds of neutral hydrogen gas in circulation around the typical spiral galaxy. These clouds are located far and near, extending well beyond the assumed boundary of the galaxy.

Fig. 14.3. This spiral galaxy M 101 is seen face-on with its spiral arms clearly visible. Credit: *NASA/Hubble Heritage.*

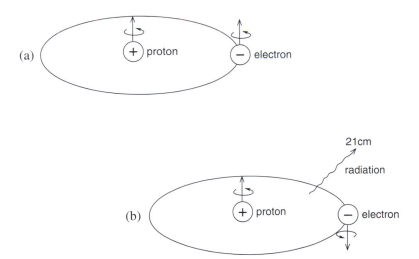

Fig. 14.4. The electron orbiting a spinning proton in the hydrogen atom may spin in two possible ways. It would spin in a direction parallel to that of the spin of the proton as in (a), or in the opposite direction as in (b). The energy of the system in the state (a) is greater than that in the state (b). Since a system tends to make a transition from a higher energy level to the lower energy level, the atomic electron sometimes spontaneously changes over from (a) to (b). In this process, the energy lost by the system appears as a photon of wavelength 21 centimetres.

From the early days of radio astronomy it was recognized that hydrogen gas in its atomic state radiates radio waves at a wavelength of 21 centimetres. Radio waves of this wavelength are emitted when the spinning orbiting electron in the hydrogen atom changes its state. From the state in which it spun parallel to the spin of the proton it changes to a state where it now spins in the opposite direction (*see* Figure 14.4). This process results in a state of lower energy and so the energy lost appears as a radio wave of 21 centimetres in length. So if we see a cloud of gas radiating at this wavelength, we can be pretty sure that it is made of neutral hydrogen gas. However, if we find that the wavelength of the radiation is *slightly longer* than 21 centimetres, say about a tenth of a percent longer, what can we conclude?

Using the Doppler effect described in Chapter 6 we would be entitled to argue that the cloud of neutral hydrogen is moving away from us with a speed of a tenth of a percent of the speed of light, that is, with the speed of 300 kilometres per second. Similarly if we see a cloud radiating at a wavelength *shorter* than the 21 centimetres by a fraction of a tenth of a percent, then we would argue that it is moving towards us with a speed of 300 kilometres per second.

In the 1960s and 70s radio astronomers were able to measure speeds of such clouds and relate them to the galaxy around which they might be moving. We may use here the analogy of the planets moving around the Sun in our own planetary system. We know from measurements of speeds of these planets that *the further they are from the Sun the slower they orbit*. For example, Mercury, the nearest planet, has an orbiting speed of around 48 kilometres per second, whereas for the most distant planet, Pluto, the speed is less than 5 kilometres per second. Applying this analogy, astronomers expected the clouds further and further away from the Galaxy to have rotational speeds smaller and smaller.

They were in for a surprise. The speeds did not seem to be dropping off; rather they stayed constant over a very long range. The figure shows results for several galaxies. As they remained flat over a long distance, these rotation curves came to be known as 'flat rotation curves'.

To resolve this mysterious behaviour, let us go back to the Solar System example. There the speed drops off because we know that the planets are moving under the attraction of the Sun and this attraction drops off as one moves away from the attracting mass. There is a definitive formula which tells us how the rotational speed of a planet should drop off with distance from the Sun. The speeds of all planets from Mercury to Pluto follow this rule. Indeed this was the classic discovery of Johannes Kepler in the early seventeenth century for which Newton's law of gravitation provided the mathematical explanation. As we mentioned in Chapter 3, the Newtonian law since then

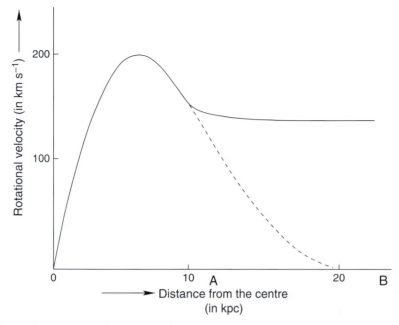

Fig. 14.5. The rotational speeds of neutral hydrogen clouds plotted at different distances from the centre of a spiral galaxy remain approximately constant. A typical behaviour found for a number of galaxies is shown by the continuous line. The behaviour expected from the Newtonian law of gravitation is indicated by the dotted line. Why do the rotation curves remain flat, i.e., parallel to the distance axis? This is the question!

has been successfully put to the test for planets, satellites and stars in the Milky Way. But the same law applied to the neutral hydrogen clouds attracted by the Galaxy, does not seem to work. Why not? Even the more sophisticated Einstein's theory of gravity fares no better.

Dark matter

Whenever there is a conflict between a well-established law and observations, two possible courses of action suggest themselves:

I. Reexamine the observations in case something crucial has been missed out.
II. Change the law for something deeper and more subtle.

The latter option calls for a fundamental rethinking and is usually put off as a last resort, especially if the existing law or paradigm has served well until this particular discrepancy. The Newtonian laws of motion and gravitation have served so well, as has Einstein's theory of relativity, that replacing or modifying them would be a high price to pay. Nevertheless there have been attempts to modify them with a view to understanding the flat rotation curves. The so-called *Modified Newtonian Dynamics* (MOND) proposed by Mordekai Milgrom is perhaps the most talked about attempt in this direction, although it is very much a minority view. In the MOND theory the second of Newton's three laws of motions is modified when applied to objects with very small accelerations.

What the majority of physicists and astrophysicists would like to follow is the first alternative. This involves admitting that our observations of galaxies are incomplete and that there is invisible matter present which extends *well beyond* the visible part of the galaxy. In terms of distances, we can argue for our own Milky Way like this. The visible matter made of stars, dust and gas may extend over a disc of radius 15 kiloparsecs. However, the *dark* matter is expected to be present well beyond this radius. It is because of this extra matter that the gravitational influence of the Galaxy extends much further, and so the rotation speeds of neutral hydrogen clouds extend without attenuation out to distances of 50 kiloparsecs or beyond.

What would the dark matter be made of ? Black holes? Since these are very efficient in holding back light, this alternative suggests itself. We could have black holes formed from remnants of massive stars that stopped shining after their nuclear fuel stocks were spent. A second possibility could be planet-like objects that are not self-luminous. Any object with mass not exceeding around a tenth of a solar mass cannot shine on its own because its core temperature is not high enough to ignite a nuclear reactor. Such objects are called *Brown Dwarfs* and these would not be seen by normal telescopes. These are examples of 'conventional' types of dark matter. Given the hypothesis that dark matter exists, these are the options one may think of in the first instance.

However, there are other more esoteric options, which are lumped together under a class called *Non-Baryonic Dark Matter* (NBDM). These options are, by definition, made of particles that *do not form parts of atomic nuclei*. Atomic nuclei contain neutrons and protons which are called *baryons* and almost all matter we see in the universe consists of these as well as light particles like electrons and neutrinos. Thus the masses of black holes or brown dwarfs are mainly made up of baryons. As yet there have been no particles discovered by high energy physicists that could be classified as NBDM. In the 1980s the possibility that neutrinos may have a fair amount of mass (corresponding to an energy of up to about 30 electron volt) had raised the option that dark matter may be accounted for by neutrinos. However, those indications have gone. Although neutrinos may have mass, it would still be far too small to explain the dark matter in galaxies. We will return to this issue of what the dark matter might be made of later, after we have described another line of evidence for dark matter on an even grander scale.

Clusters of galaxies

In the adjoining figure we show a picture of a cluster of galaxies. A typical cluster contains several hundred galaxies and they are all moving randomly within the cluster. These random motions are of the order of 250–500 kilometres per second. *These motions are over and above those arising from the expansion of the universe.* Thus a typical cluster partakes in the overall expansion process, and additionally has galaxies moving within it at random speeds.

If we assume a cluster is an isolated dynamical system of many bodies which have been moving under one-another's gravitational attraction for a long enough time to settle down to some steady state, then we can deduce a simple result from Newton's laws of motion and gravitation. It is that the energy of motion, the so-called *kinetic energy* of all moving galaxies is comparable in

Fig. 14.6. The Coma cluster contains nearly a thousand galaxies. The energy of motion of all of them seems to greatly exceed their gravitational energy, leading to the hypothesis of dark matter. Credit: *Atlas Image courtesy of 2MASS/UMass/IPAC-Caltech/NASA/NSF.*

magnitude to their total gravitational potential energy. This is known as the *virial theorem*. So if we estimate the two energies for clusters, we can verify if the virial theorem does apply to them.

For most clusters it does not. The energy residing in the motion of the member galaxies is much higher than the energy residing in gravitational attraction. The discrepancy is large enough to make one think. One possible conclusion can be that the clusters have not yet had time to settle down and so the virial theorem does not apply to them. This could happen if the cluster is expanding or contracting all round. The Armenian astrophysicist Viktor Ambartsumian had concluded back in the early 1960s that the clusters are expanding, having been created in an explosion. Based on his assessment of the data, Ambartsumian concluded that the clusters are examples of explosive creation of matter. We will return to this conclusion in the next chapter.

The majority view, however, is different. The view is that the clusters have indeed settled down to an equilibrium state and the reason we have a deficiency of gravitational energy is because we are not able to *see* all the matter present in the cluster. Suppose there is a lot of dark matter within the cluster which is not moving fast. Such matter will not contribute much kinetic energy, but would give rise to large gravitational energy by virtue of its mass. This is why we notice a deficiency of gravitational energy.

This argument has therefore suggested to the theoreticians that they can add as much dark mass as they need to make up the energy deficiency. The amount of dark matter to be added this way far exceeds the visible matter. Whereas in the case of rotation curves of spiral galaxies the ratio of dark to visible matter may be around 1 to 3 or so, in the case of the clusters the ratio may go up as high as 10 to 1 or even more.

What is dark matter made of?

So, now we come back to the question posed earlier. . .what is such dark matter made of? Even though it is not seen, we can argue for the options like black holes or brown dwarfs. However, there are problems with these options. First one has to argue for a physical scenario that led to so much of matter being in this form. This may or may not be a very difficult problem. . .with sufficient ingenuity, the theoretician may come up with a plausible scenario. But theoreticians of another breed, those who adhere to the big-bang model, object to this possibility.

The big-bang theorists would be worried if so much dark matter existed in these relatively normal forms. For these forms are all made of baryonic dark matter (BDM). If there were so much BDM around, a difficulty arises with the big-bang scenario of how light nuclei, especially deuterium, were made. In the process of primordial nucleosynthesis first proposed by George Gamow, and later worked on by several other astrophysicists, one crucial conclusion described in Chapter 11 was that, provided the density of baryons exceeded a critical limit, *practically no deuterium would be made*. And if we begin to allow all or most of dark matter in clusters and galaxies to be baryonic, that critical limit would certainly be exceeded.

In fact, a difficulty of even greater magnitude awaited the big-bang theorist when he resorted to the inflationary scenario described in Chapter 13. If inflation did happen, it would leave the universe with a density very close to the critical density. If all this matter were normal baryonic matter, its density would be 25–30 times higher than the limit tolerated by the deuterium synthesis process.

So the conventional big-bang theory runs into a serious problem. If it allows inflation, it runs foul of deuterium production in the primordial nucleosynthesis. It also ends up with far more dark matter than the evidence from galaxies and clusters suggests. This latter difficulty can be resolved by supposing that there exists dark matter not only inside clusters but also in the space between them. However, the first problem was more serious. To find a way out therefore, big-bang cosmologists have supposed that the bulk of dark matter in or out of the clusters is *non-baryonic*. We have already mentioned this option previously. The non-baryonic dark matter (NBDM) is an esoteric option which has to be adopted because there is *no other alternative for survival of the big-bang nucleosynthesis scenario*. An alternative name given to such an NBDM particle is 'weakly interacting massive particle' or a WIMP!

Why do we call NBDM esoteric? Because there has been no laboratory demonstration of it. Nor has it been detected in any astronomical scenario. Rather, the theoretical possibilities for such matter come from the as yet untested theories of very high energy particles. As we mentioned in the last chapter, man-made accelerators do not reach these kinds of energies. So what we are effectively asked to accept is that the bulk of matter in the universe is of this kind, far exceeding the normal kind of matter that we are familiar with in nature.

The reader may once again recall here the paradigm of 'anti-Earth' of the Pythagoreans (*see* Chapter 2). The anti-Earth was not seen; but it had to be accepted because it successfully hid the central fire that had also not been seen.

But to proceed further, once one accepts the paradigm of NBDM, then one's imagination leads to further speculations. Is the NBDM hot or cold? Big-bang cosmologists talk of two kinds of non-baryonic dark matter. The hot dark matter (HDM) consists of particles which were in close interaction with normal baryonic matter in the early times, and when their interaction ceased, they were still moving fast, with speeds near that of light. The other species of dark matter is the cold dark matter (CDM) which is made of particles which had slowed down to almost the state of rest by the time they ceased to interact with ordinary (baryonic) matter.

The particle which comes nearest to a realistic example of an HDM is the massive neutrino. The neutrino interacts with ordinary matter very weakly, and theory tells us that when the universe was less than a second old, the neutrinos stopped interacting with other light particles like the electrons in any significant way. At that stage the typical neutrino was still moving very fast and so the neutrino is a particle of the HDM type. Depending on its mass (still not reliably determined) it would slow down; but this would have happened much later.

What are examples of CDM? These would depend on a typical high energy particle theory. The menu of possible particles is quite extensive, the more popular ones being the photino, gravitino, the axion, etc. Rather than go for a specific particle, the cosmologist proceeds in the opposite way. He asks, what should such a particle do in order to be consistent with the requirements of the cosmological theory? These requirements include the scenario of formation of galaxies and the tiny fluctuations observed in the microwave background described in Chapter 12. In this sense, the approach has no scientifically predictive power. It only asks for overall consistency in the picture. Since no CDM particle has been found, the issue is wide open for speculation about what it should be like. Indeed, since the late 1980s, CDM has held centre stage in the drama of dark matter.

However, speculations in modern cosmology do not end there.

Dark energy

We recall that in the early stages of cosmology, Einstein had introduced cosmological force in his equations, to obtain the mathematical model of a static universe. Later when he discovered that observations favoured an expanding universe and that his original equations did yield expanding models, he more or less abandoned this extra force.

Cosmologists have since had a love–hate relationship with cosmological force. Whenever they feel that their models are threatened by new observations they invoke the force, perhaps with reluctance, only to abandon it if later it is discovered that the observations were not threatening after all. As mentioned in Chapter 7, the intensity of this force is typified by a constant often denoted by λ or Λ. The constant in today's universe is very small and this indicates that the force of repulsion implied by it is very small on the terrestrial, stellar or galactic scale. However, on the scale of the universe as a whole, it is significant. A positive Λ means the force is of repulsion and on a large scale it makes the universe accelerate. Is the universe really accelerating?

In Chapter 10 we described the extensive work by Allan Sandage to check on this very fact. In the 1960s and 1970s, the results of studies of distant galaxies indicated that the universe is *decelerating*, that is, its rate of expansion is slowing down. At the time, the Friedmann models without the Λ-term were popular and these indicated the same conclusion. The only model that stood apart was the steady-state model which implied that the universe is *accelerating*. Later this test fell into disuse as it was realized that there were several imponderables, observational errors that made any definitive conclusion impossible.

However, the test was revived in the 1990s when it became possible to make dedicated studies of exploding stars, called *supernovae*, lying in distant galaxies.

Fig. 14.7. The bright dot seen in the image of a galaxy is identified as a Type Ia supernova. Credit: *NASA*.

A particular class of supernovae, called Type Ia supernovae, seemed to have the property that they provided a standard candle for measuring galactic distances. Let us first try to understand what this statement means.

The figure shows the brightness variation of Type Ia supernovae. Typically, a supernova of this type represents a highly compact star blowing up as it loses its internal equilibrium. The intensity of the star shoots up after the explosion and it reaches a peak in luminosity within a few days. The typical light curve of such a supernova is shown below.

The important thing to note is that the supernova becomes very bright and may outshine the entire galaxy in which it is housed, but for a few days. The peak luminosity therefore makes it easy to spot a supernova even if it is located in a very distant galaxy. Also, it seems that the maximum brightness attained by the star is more or less the same from one Type Ia supernova to another. So we can use the method of measuring distances outlined in Chapter 5 to estimate the distances of galaxies in which the supernovae are located. The fainter the supernova the further away it is, as per the rule that candles further away look dimmer. The fact that the peak intensity for all Type Ia supernovae is the same is called the 'standard candle hypothesis'.

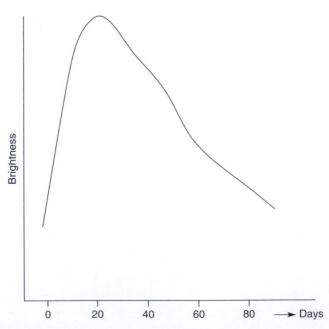

Fig. 14.8. Light curve of a Type Ia supernova. Notice the rise in intensity soon after the star explodes, followed by a fall over several weeks. It is assumed that the peak intensity remains more or less the same for all such supernovae.

A 'Supernova Cosmology Watch' programme was set up to observe and record any such sudden eruptions in galaxies with redshifts ranging up to around 1–1.5. These redshifts are higher than those of galaxies used by Sandage in his earlier studies...those went up to around 0.5. Thus we are in principle able to sample a more remote part of the universe with the help of supernovae.

The method is then to look at supernovae at different distances and see how their redshifts change with distance. Redshifts are obtained, as explained in Chapter 5, by studying the spectra of galaxies, while distances are estimated by using the standard candle of Type Ia supernovae. Broadly we expect that if the universe is decelerating the distances will increase with redshift more slowly than if the universe were accelerating.

If the observers hoped to find a confirmation of the earlier results that the universe is decelerating, they were in for a disappointment. The distances as estimated from supernova standard candle seemed to increase faster with redshift than allowed by any decelerating model. Rather the indications were that *the universe seems to be accelerating!*

At this stage it would have been fair on the part of observers to have acknowledged that the conclusion in favour of an accelerating universe was

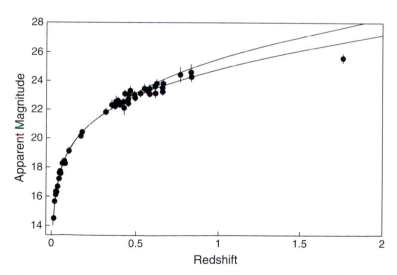

Fig. 14.9. The curve shows the apparent magnitudes (plotted on the horizontal axis) of supernovae of Type Ia at their peak intensity and at different redshifts (on the vertical axis). It is argued that the supernovae at large redshifts are dimmer than expected if the universe were decelerating. This is seen from the fact that the typical theoretical curve as shown in the lower line in the above figure lies below the plotted points. To fit the data, big-bang cosmologists are forced to assume that the universe is accelerating. This assumption raises the curve to the upper level.

consistent with the prediction of the steady-state universe. Even though in the 1990s, the steady-state universe was no longer in serious contention, a note of this historical fact should have been normal practice in science. However, the result was simply announced as favouring the standard big-bang model with a non-zero cosmological constant.

That this was a volte-face on the part of the big-bang establishment can be seen from the fact that as late as 1997, the general belief was that there is no cosmological constant and that the universe is decelerating. While changing the model so significantly from what had been previously in vogue, an acknowledgement should have been made that such a change was being forced on the theory by observations. That the present approach has no predictive value is seen from the circumstance that today's observers ask the following question: *What value of the cosmological constant will give a good fit to what is observed?*

Like good salesmen for the inflationary hypothesis, the cosmologists announced this finding as confirming the inflationary paradigm by arguing that the results bore support for the conclusion that the universe is flat, i.e., with $\Omega = 1$. What was not emphasized was that the data gave best fit to the value $\Omega = 1.3$.

According to current wisdom, the density parameter Ω is made up of three components: (1) visible (baryonic) matter, (2) cold dark matter (CDM) and (3) dark energy. We have already elaborated upon the first two of these. The third component is related to Λ, the magnitude of the cosmological constant. After studying the supernova results and also the fluctuations of the microwave background described in Chapter 12, cosmologists have come to the conclusion that the contribution to Ω from these three components can be quantified quite precisely as follows: (1) the contribution of baryonic matter is 4%, (2) the contribution of CDM is 23% and (3) the contribution of dark energy is 73%. This conclusion is referred to as 'precision cosmology'. The impression is created that with such precise details the long-standing cosmological question: 'What is our universe made of?' is finally answered.

If these 'precise' values are to be believed, then cosmologists are telling us that the only form of matter and energy that astronomers see, occupies only 4% of all matter-energy in the universe. The remaining 96% is made of esoteric dark matter and the even more esoteric dark energy. Like the emperor, the universe apparently requires invisible clothes!

What is dark energy?

The revival of the cosmological constant has raised fresh issues. If we assume, following Einstein, that the cosmological constant has been constant at all times, then we run into a new difficulty vis-à-vis inflation. The universe was driven to inflate because of the extra energy it obtained from a phase transition. That energy is very similar to the dark energy and the inflation of the universe was very similar to that in the classic 1917 model of de Sitter (see Chapter 7). Thus there was an effective cosmological constant that drove the inflationary universe. Only the time scale for inflation was very short and so the corresponding cosmological constant was very large, compared to its present value. How large? It was large by a factor 10^{108}, that is, by a factor represented by a number which has 108 zeros after 1. So prima facie one is forced to conclude that after the inflation was over, the extra energy almost disappeared, leaving behind an extremely tiny fraction of the order of one part in this large number! Further, this leftover has to be very finely tuned, otherwise, the whole expansion of the universe would go astray.

This is ironical, since the one reason for invoking inflation was to avoid fine tuning of precisely this nature. The flatness problem required fine tuning and to avoid that inflation was proposed. Now it appears that inflation brought its own fine tuning to an even greater degree! To avoid this problem one needs to have a dynamical mechanism which would reduce the cosmological constant from its initial very large value to what is required today.

Naturally theoreticians are busy trying to provide a respectable mathematical model to patch up this defect. At the time of writing there is no satisfactory model achieving this, although the confidence with which the existence of dark energy is believed exceeds confidence in Newtonian gravitation.

Concluding remarks

We can summarize as follows. It is clear that the important observations of flat rotation curves of galaxies opened up the Pandora's box of dark matter. The evidence for dark matter is certainly there if one continues to have faith in the laws of Newton and Einstein. However, how much dark matter is really warranted? If one is not prejudiced by belief in inflation then one need not have $\Omega = 1$. One can manage with much less matter. Can it all be baryonic as our experience of the rest of astronomy would have us believe? If you are not committed to the notion of primordial nucleosynthesis, then the answer is 'yes'. But if one is firmly of the view that inflation did take place and that light nuclei were made in a primordial nucleosynthetic process, then one is driven to postulating that a lot of dark matter is esoteric and non-baryonic.

Coming to dark energy, the major argument in its favour rests on inflation and the observations of distant supernovae. But there too the chain of reasoning may have glitches. Are we sure that the standard candle hypothesis is valid? If there is significant variation in the peak intensity of light from Type Ia supernovae, then the distance measurement on which the test rests is not so reliable. When we infer the distance of a supernova from its observed faintness, we ignore the presence of any absorbing intergalactic dust. Our knowledge of the intergalactic medium is still very primitive, and by ignoring intergalactic dust in estimating distances, we may be committing the same

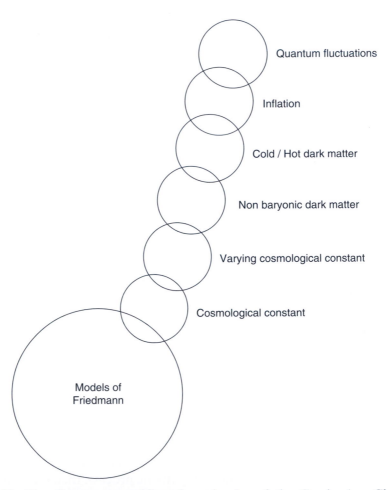

Epicycles of modern cosmology

Quantum fluctuations

Inflation

Cold / Hot dark matter

Non baryonic dark matter

Varying cosmological constant

Cosmological constant

Models of
Friedmann

Fig. 14.10. The diagram resembling the epicycles of the Greeks (*see* Chapter 2) symbolically indicates the number of different assumptions that the big-bang theory has had to make in order to survive. There is no independent evidence sustaining these assumptions. These assumptions include, dark energy, non-baryonic dark matter, inflation, biasing, etc.

error that galactic astronomers committed a century ago when they were estimating stellar distances without knowledge of interstellar dust. Intergalactic dust will make a supernova look dimmer than in the absence of dust and so if we ignore this effect we will overestimate the distance of a supernova and this error will grow the further away the supernova is. So instead of the cosmological constant causing an accelerated universe in which all distances are enhanced, it may be the absorption by dust that makes high redshift supernovae look dimmer.

Even if we discount the dust alternative and stick with the accelerating universe, we find that data do not really fit the simple model in which a constant Λ accelerates the universe. One needs a variable Λ, thus making the hypothesis messier. For, more recent evidence apparently points to acceleration over a limited period. Thus theoreticians are getting lost in more and more complex models of dark energy, which have no predictive power.

As we discussed at the end of Chapter 12, the interpretation of cosmic microwave background fluctuations also rests on a number of assumptions, none of them resting on direct facts but on untested extrapolations of known theories. Indeed, despite claims made by big-bang cosmologists, that they have arrived at a consensus on what the universe is like, and that they have its parameters precisely determined, we see that their entire façade rests less on direct facts than on a chain of hypotheses reminiscent of the epicycles Greeks had built to support their theory of planetary motions.

Given this history of standard big-bang cosmology, and its current bag of speculations, we feel that there are certainly no firm grounds for assuming that it provides a factual account of the real universe. It may well turn out that all of today's speculative elements will be borne out by facts at a later stage. If that happens we can accord full support to this cosmology. Until then, however, a sceptic may be justified in thinking that reality may lie elsewhere.

With that sceptical view we now consider in the following chapter, an alternative scenario that does not involve the big bang.

15

An alternative cosmology

Did the big bang really happen?

In the preceding chapters we have been somewhat sceptical about the standard model of the origin and evolution of the universe. According to this model the universe had gone through the following stages:

I. First there was a singular origin (the big bang), involving a quantum era of uncertainty where extremely small scales of space and time are invoked. They require a new approach to the study of gravity. This era lasted up to a time of the order of 10^{-43} second. Tiny though this period was, it was important from the point of the origin of large-scale structure. This was the period when primordial seeds of density fluctuations that later grew into galaxies, were sown.

II. Following this there was an era of grand unification of the rest of the three basic interactions, that lasted up to time of the order of 10^{-37} second. So far these are only words, there is no real physical theory behind them.

III. An inflationary era then followed, during which the universe inflated enormously in size and very fast. This era may have lasted for around 10^{-35} second. Inflation is believed to have played a key role in controlling the distribution of matter inhomogeneities that grew into galaxies and clusters later.

IV. In due course the universe passed through a period when it managed to have a lot more particles of matter than antimatter, so that the antimatter was annihilated finally ending with a brew of baryons (neutrons and protons) and leptons (electrons, positrons and various types of neutrinos) as well as considerably more non-baryonic dark matter.

V. One second after the big bang, the process of formation of light atomic nuclei started, and it ended about three minutes after the big bang, by which time around 24% of matter was converted into helium with tiny amounts of deuterium, lithium and a few isotopes of atomic weight not exceeding 8.

VI. Around a hundred thousand years after the big bang, the radiation that was until then closely interacting with matter, decoupled from it and began to travel freely through space. This happened because the electrons that were scattering the radiation became trapped in atoms and were no longer able to scatter the radiation. After this epoch observational astronomy in the conventional sense became possible.

VII. In the meantime the formation of large-scale structure was going on through gravitational interactions amongst primordial lumps of matter that were steadily growing. Subject to the influences of cold dark matter and dark energy, the lumps grew to the presently observed size of large-scale structure.

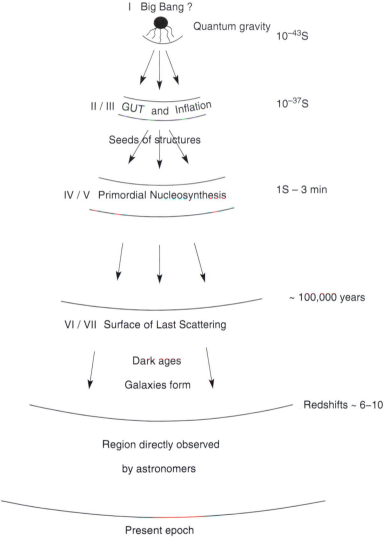

Fig. 15.1. A schematic view of the standard model of the universe. Direct observations are limited to the last scattering surface and in fact up to today extend to a region far smaller. The extrapolations of known physics that relate to the very early epochs have no independent supporting evidence.

This entire scenario is depicted in the above figure showing the various stages of what happened when. There are still large gaps in our picture, for example, the gap between stages 6 and 7 is sometimes called the 'dark ages' because we do not have any specific scenario of how the universe evolved from an age of 100 000 years to the epoch when the earliest galaxies formed.

Thus we have a 'standard' model which claims to have a concordance of various 'observational checks' and to have a rather precisely determined set of parameters describing the universe. It is reasonable to ask what is the direct observational evidence for all these steps, and also how soundly are the physical concepts behind the theoretical claims tested?

Although this model is generally believed, there is no *direct* observational evidence to support it. As far as direct observational checks are concerned, we are limited to the distant galaxies that can be detected and have their redshifts measured. Even with the most sophisticated telescope with very large size, this limit into the foreseeable future is about redshift 10. That is, we are able to see what the universe was like when its linear scale was 1/11 of its present value. The stages 1−6 came earlier than that. Stage 6 in fact tells us that before it was over, radiation was so effectively scattered that it was useless as far as any information from the original sources is concerned. It is like trying to see through a thick fog. We are like a theatre audience trying to figure out what is happening on the stage when the main curtain is down. We cannot see through the curtain, but we may try to surmise what may be going on by observing the tiny disturbances of this screen. We may be correct in our conjectures and inferences, or we may be wrong, but we have no way of telling for sure.

The reason for a lack of confidence in the correctness of such surmising comes from the fact that the physical theory used for inferring what went on in stages 1−4 is untested in the laboratory. In fact it is not unique, and particle physicists are still uncertain about its final format. Moreover, the sequence of events that is believed to have taken place in the scenario is *unrepeatable*. It happened only once and could not be observed again. But in science, the credibility of a result rests on more and more experiments being performed to demonstrate its reality. This facility is denied in this type of cosmology.

We have stated this before, and feel that the issue is important enough to be stated again. We can draw comparisons from two separate fields of science so as to contrast the present situation with what happens there. First look at the physics of stars. There we have a physical theory based on known and experimentally tested areas of science. The interior of a star can be analyzed and deductions can be made as to how much luminosity a star of specified mass will have. This checks well with the Sun. But the Sun is only one example. There are hundreds of stars of varying masses which we can observe for which the theoretical mass–luminosity relationship can be tested. Thus we may look on all these stars as separate trials for testing the theory. And it comes out correct. In Figure 15.2 we show the situation.

Our second example comes from quantum mechanics. This is the branch of physics which deals with the microscopic world of atoms and molecules and the nuclei of atoms. It could be argued that we are not able to observe these systems directly and so the area is speculative. Although recent advances in nanotechnology have made it possible to counter this criticism, long before this, a large number of experiments performed repeatedly under different controls had already demonstrated the validity of quantum mechanics.

In view of this situation why is it that the vast majority of astronomers and cosmologists – and through them the media and the general public have become convinced that the big-bang hypothesis is the correct one?

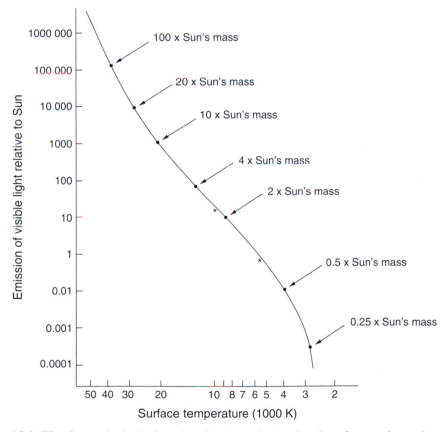

Fig. 15.2. The theoretical relation showing how the luminosity of a star depends on its mass when the star is on the main sequence, has been verified for many stars of different masses.

In our view this is because of the impact that the first major discovery – the redshift–apparent magnitude relation found by Hubble and Humason – had on the community and the conclusion drawn from it. The conclusion was that the universe is expanding, and this was compatible with the solution to Einstein's equations found by Friedmann in 1922, and by Lemaître in 1927. It was seen as confirmation of Einstein's theory of gravity. By 1929–30 it was concluded that the universe must be expanding, and it appeared obvious that the universe must have been smaller at earlier times. This led to the idea that the universe began as Lemaître put it as a 'primeval atom'.

It took about twenty years before any real physics could be put into the problem, but by the late 1940s enough was known to believe that when it was very young the universe must have been made up of protons, neutrons, electrons, neutrinos and radiation.

This general picture of an evolving universe with a beginning was satisfactory to most philosophers and Western theologians, since it is generally compatible with the idea of a beginning in Western culture.

The belief in an early dense phase in the universe was given a further boost in the early 1950s when George Gamow and his colleagues became convinced that there would be an early fireball associated with this phase, and this would expand indefinitely maintaining its black body form. The fact that there was observational evidence for such radiation as early as 1941 was not known to Gamow, or any cosmologists. But the Princeton group reworked the whole process and set out to detect the radiation. When Penzias and Wilson found a radiation field in 1965, this was seen as confirmation of Gamow's ideas and from that time onward it was generally accepted that observational evidence overwhelmingly established the correctness of the evolving (big-bang) universe model.

In retrospect, it is clear that we have to go back to the interpretation of the expansion result and ask whether there are alternative explanations.

In 1947 Tommy Gold argued that while the universe is expanding it may not be changing, and thus there may never have been an initial state. It was Gold's intuition which led to the formulation of the classical steady-state theory of Bondi and Gold, and of Hoyle. We have explained elsewhere why this classical model is in conflict with the observations (Chapter 12). However, it turns out that a model in which the universe repeatedly goes through cycles of expansion and contraction, but never reaches extreme conditions in its contraction phase is compatible with all of the observations. This is the quasi-steady-state model (a cyclic universe) which we developed with Fred Hoyle in the 1990s.

Before describing this we discuss the strongest evidence in favour of the standard model.

First, the observed abundances of light nuclei are consistent with what is produced in the primordial nucleosynthesis process. It is argued that there is no alternative way of making these nuclei, such as in stars. We will come back to this argument later. However, as mentioned in Chapter 11, the success of the theory of production of light nuclei in the primordial process is contingent on a very finely tuned relationship between the density of baryons and the ambient temperature. In the case of deuterium, this places a very stringent upper bound on how many baryons should be found per unit volume at present. This is why when dark matter density is taken into account, we need to postulate the existence of non-baryonic dark matter (NBDM). The presence of NBDM is thus inferred not from any direct finding in the universe but by the requirement of survival of the standard model.

The most powerful evidence claimed for the standard model is that from the cosmic microwave background. We have discussed this in Chapter 12. The data from COBE in 1990 provided a comprehensive spectrum of the background and showed it to be very close to the black-body form. Such a spectrum is predicted by the standard model for relic radiation from the big bang, and the fit between theory and observation is remarkably good. Moreover subsequent studies of the fluctuations of the background by many satellite and ground-based observations starting with COBE in 1992 and ending with WMAP in 2003 have been explained well on the basis of the standard picture. However, technically speaking, the observations relate to the radiation found *today*. They do not measure radiation present at the so-called last scattering surface, when the universe was supposedly around a few 100 000 years old (stage 6 above). Any linking with that epoch has to rest on theoretical extrapolation from *then* to *now*.

Comparatively recent observations of high redshift supernovae are used to infer the existence of dark energy. As was mentioned in the last chapter, elaborate modelling of how dark energy behaves at different epochs is needed in order to reconcile theory with observations, *there being no independent theoretical justification that dark energy must behave in this fashion*. Also, not enough attention has been paid to other effects, such as the breakdown of the hypothesis of standard candles, the effects of gravitational lensing and also likely absorption of light by intergalactic dust. All of these may lead to large corrections in supernova brightness.

Given this state of affairs, a sceptic may be justified in asking whether the cosmologist should place all his eggs in the big-bang basket. Given the history of the subject is it not possible to look for an entirely different explanation *that does not require a big-bang origin, that does not require inflation* and *which relies more on direct observations than on large theoretical extrapolations of theory*?

We explore one alternative cosmological model which we favour, in this chapter.

The quasi-steady-state cosmology

In Chapter 9 we described steady-state cosmology as an example of a universe without a beginning and without an end; a universe that never had a singular state. This model had provided a possible alternative to the big-bang cosmology, at least until the mid 1960s, when it stumbled, on evidence for the microwave background and the abundances of light nuclei. This model was revived in a modified form in 1993 by Fred Hoyle and the two of us, in the form now known as *quasi-steady-state cosmology* (QSSC in brief hereafter). It retained many of the features of the steady-state theory but had the added feature that over and above a long-term steady expansion, the universe oscillates with a shorter period. The adjoining figure demonstrates this feature.

In this picture, the characteristic long-term time scale of expansion is 1000 billion years, while each oscillatory cycle has a period of 50 billion years. Notice that, the universe never had its scale factor going down to zero. Further, although it expands and contracts over a cycle of 50 billion years, on the longer term, there is an overall expansion. (We encounter similar behaviour with the price of fruit and vegetables, with prices rising and falling seasonally every year, although over a longer term of 5–6 years, there is an overall rise in prices.) Why and how does such behaviour arise?

The clue to this lies in the phenomenon of matter creation. In the case of the steady-state theory, we encountered the so-called *C*-field, which has negative energy and which grows in strength whenever it creates matter. The steady-state universe is driven by this field and it expands because of the gravitational

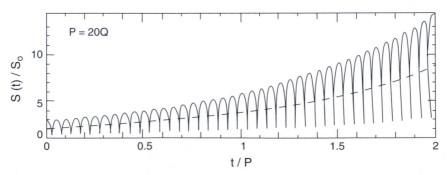

Fig. 15.3. The scale factor of the quasi-steady-state cosmology is shown above at different epochs. The universe expands and contracts in a typical cycle with a time period Q = 50 billion years or so. On an even longer term of around P = 1000 billion years, the universe has a long-term steady expansion.

repulsion produced by the *C*-field. This expansion is steady and the same at all points of the universe, because matter is created *at a steady rate homogeneously*.

A closer reexamination of the creation process first made in the 1990s showed that the condition for creation of matter is linked with the mass of the particle created. Theory showed that the characteristic particle to be created is the so-called *Planck Particle* (PP in brief), which is the fundamental classical particle indicated by a quantum theory of gravity. Although we are still far from understanding what a quantum theory of gravity is like, the result that physicists would agree on is that a particle whose mass is made up of the three fundamental constants of nature, namely the gravitational constant G (which measures the strength of the gravitational force), the speed of light c (which plays a fundamental role in the special theory of relativity) and the Planck constant h (basic to quantum theory) should play a key role in such a theory. This is the Planck particle, and its mass is about a few micrograms. This is the particle that is first created in a creation process. Thereafter it decays into a large number of smaller particles including baryons (like neutrons and protons) and leptons (like electrons and neutrinos).

It turns out that the overall level of the *C*-field everywhere is not high enough for such a particle to be created. Thus we can say that the threshold for creation is so high that, barring isolated exceptions, most regions in the universe would not see matter being created. What and where are those exceptions?

They are massive objects which are very compact and dense. Near such objects the level of the *C*-field is raised to a high enough level so that creation of PPs is possible. For this to occur the strength of the gravitational field needs to be almost as high (but not quite) as that of the classical black hole. We can see this comparison in the following way. If a spherical object has mass M, then its radius in the black hole state has to be as low as twice the product of M and G divided by c^2, i.e., the product of c, with c. Only when it has shrunk to this size will its gravity have grown so intense that it pulls back light quanta (radiation) which are trying to escape.

For the Sun to become a black hole it would have to shrink to a radius of only about 3 kilometres, whereas the actual radius of the Sun is around 700 000 kilometres. Because the *C*-field is repulsive, its presence prevents a massive object from becoming a black hole. Thus if such an object begins to shrink because of its own gravity, it will ultimately slow down to a halt and bounce, *before becoming a black hole*. It is such objects, which we may call 'Near-Black Holes' (NBHs), that act as centres of creation. For, close to their surfaces the strength of the ambient *C*-field rises high enough to initiate creation.

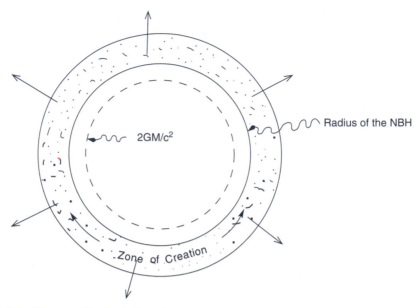

Fig. 15.4. The small shaded region surrounding a dense spherical mass has the intensity of *C*-field high enough to trigger the creation of new matter. The created matter is ejected in an explosive fashion as indicated by the outward arrows.

What happens when creation begins near a typical NBH? The process is unstable in the following way. Creation of matter increases the strength of the *C*-field and thus provides a further boost to the creation process. At the same time the growth of the negative energy reservoir of the *C*-field creates a situation where repulsion begins to dominate over attraction and the outcome in a short time is to throw out all the new matter created away from the NBH. This is the reverse of the collapsing stage which originally produced the NBH. Indeed in a sense we may call it a *white hole*. In a white hole matter moves outwards in an explosive fashion and radiation brightens up as it too moves outwards. Contrast this situation with that prevailing in a black hole which holds back all of its radiation under its strong attraction.

This explosive situation may be called a *minibang* or a *minicreation event* (MCE).

The minibang may at first sight appear to be a small-scale replica of the big bang...that is an event with very high energy and dense matter created explosively. In reality, however, there is a great deal of difference. In an MCE matter and energy are created. But the process is governed by known laws of physics, rather than given as a fait accompli in some mysterious manner as in the big bang. There are no infinities to worry about either, since all processes are finite in magnitude and subject to the same mathematical manipulation that applies to the rest of physics. But perhaps the biggest difference lies in the fact that the MCEs are observable and repeatable events, unlike the big bang which is believed to have happened only once and that too in an unobservable fashion. We will return to the observational evidence for MCEs shortly.

The theory that describes the gravitational effects of the *C-field* and matter, further tells us that there is also a cosmological constant in this picture, only *it is negative*. That is, it generates a long-range force of attraction between any two particles of matter, an attraction that increases in proportion to the distance between them. So we have two sources of attraction, the normal gravitational force and this cosmological force coming from a negative cosmological constant and the force of repulsion arising from the *C*-field. Together these forces control the expansion of the universe driven largely by the minicreation events just described.

But what is the connection between the MCEs and the expanding universe? We note first that each MCE, by virtue of its explosive nature and abundance of *C*-field, creates a zone of repulsion around itself. So if there are a number of such MCEs distributed all over the universe, there will result an overall expansion of space containing them. The rate of expansion will be related to the typical intensity of an MCE and will also depend on their overall abundance. However, instead of a steady expansion resulting from this process, what one gets is a 'quasi-steady expansion'. In this type of expansion the creation activity keeps going up and down, for the following reason.

Whenever there is expansion of space, the intensity of the *C*-field drops. When its overall level is too low, it becomes harder for an NBH to raise it to the threshold for creation of matter. So many MCEs cease to function. This results in a slowing down of expansion, and coupled with the attractive nature of the cosmological force, ultimately leads to contraction of space. But this phenomenon is temporary too, for with contraction the strength of the *C*-field grows and more and more of the MCEs that had become defunct as producers of matter now come on line, and the production of matter grows along with the production of the *C*-field. These effects result in a halt to the contraction of space and its conversion to expansion. Thus we have cycles of expansion and contraction superposed on the overall long-term expansion of the universe as described earlier.

We now relate this model to physical reality.

Explosive events in the universe

In the 1960s, the great Armenian astrophysicist Viktor Ambartsumian had suggested that there was growing evidence for pockets of explosions, which, he believed, produced new matter in the universe. These explosive phenomena

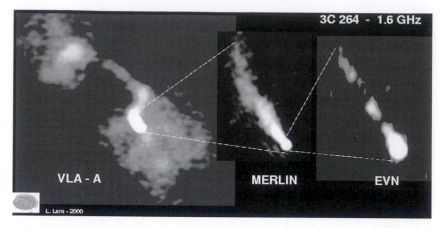

Fig. 15.5. 3C 264 is an FR-I source which contains a jet that is clearly visible in both the radio and the optical telescopes. Seen with VLA, MERLIN and EVN, different resolutions reveal radio emission on different scales: from the large extended lobes seen with the short baselines of VLA to the collimated jets that are seen at the highest resolution with EVN. Credit: *Lara L., Giovannini G., Cotton W. D., Feretti L., Venturi T.*

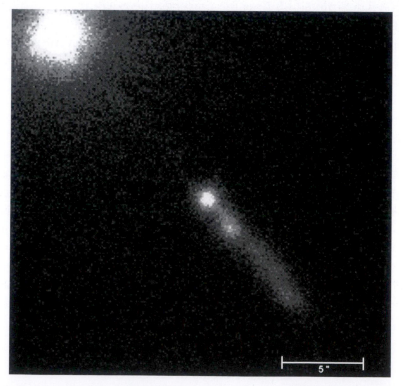

Fig. 15.6. Compared to the radio galaxy shown in the preceding photograph, here we see a much more compact source, a quasar. The quasar 3C 273 shown above was one of the first two to be so identified. It shows an optical jet. 3C 273 emits strongly in the optical, radio as well as in the X-ray wavelengths. Credit: *Chandra Observatory*.

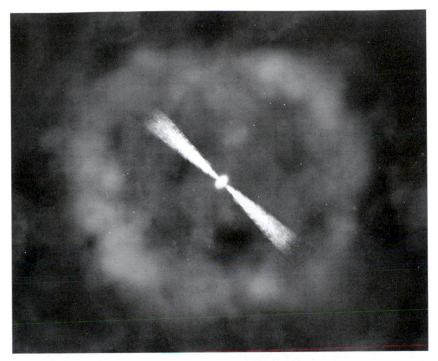

Fig. 15.7. This image of a gamma ray burst called GRB 031203 also shows a jet-like structure. Credits: *Illustration: NASA/CXC/M. Weiss.*

have now proliferated in number as well as variety. We may cite some examples.

I. We have radio sources that show plasma coming out in the form of jets in the opposite direction. Typical pictures are shown in Figures 15.5 and 15.6. The plasma is believed to hit the intergalactic medium and the fast-moving electrically charged particles resulting from this impact radiate radio waves. The central region of the source shows explosive activity and outward motions of particles.

II. The quasi-stellar objects described in Chapter 10 show evidence of production of large amounts of energy in a very compact region, energy that is ejected outwards as radiation in the form of visible light, X-rays, infrared, ultraviolet and radio waves.

III. The discovery of gamma ray sources in the form of bursts of energy suggests the emission of huge amounts of energy in an explosive fashion. We have a picture of one of these objects above.

IV. Ambartsumian had also drawn attention to the larger scale phenomena of clusters of galaxies. He did not believe that these were all systems in equilibrium. Recall when we discussed the evidence for dark matter in clusters (*see* Chapter 14), we had assumed that the clusters are in dynamical equilibrium, so we

Fig. 15.8. This Hercules cluster of galaxies may not be in dynamical equilibrium. Credit: *NOAO/AURA/NSF*.

expected that the kinetic and gravitational energies were comparable in magnitude. Reality shows in most cases that the kinetic energy exceeds the gravitational energy. Ambartsumian took this to mean that in some cases the clusters are *not in equilibrium*. Rather he interpreted evidence to imply that these clusters are expanding and inferred that they had originated in explosions.

The QSSC was inspired by Ambartsumian's ideas, the new data and by the original steady-state theory. We now look at how the theory copes with the observed features of the universe. In particular, can it explain the origin of the cosmic microwave background and also the origin of light nuclei? These were the aspects in which the original steady-state theory had been found wanting. We consider the origin of nuclei first.

Light nuclear abundances

As was mentioned in Chapter 11, it is generally assumed that the production of chemical elements in the universe arises mainly from two processes. In the early big-bang universe, during the era beginning from a few seconds to about three minutes, the light atomic nuclei were made. Heavier nuclei were made in stars. This belief was challenged in 1998 by Geoffrey Burbidge and Fred Hoyle. To understand their argument we first return to the basic structure of atomic nuclei.

The typical atomic nucleus contains charged particles called *protons* and slightly heavier neutral particles called *neutrons*. The chemical properties of the atom depend on the number of protons it has. Two nuclei containing the same number of protons but different numbers of neutrons will have the same chemical properties. Such nuclei are called *isotopes* of the same chemical element. The hydrogen atom has one proton in the 'normal' atom and no neutron. A 'heavier' version of hydrogen exists in nature whose nucleus, called *deuterium*, contains one proton and one neutron. A normal water molecule is made up of two atoms of normal hydrogen and one atom of oxygen, giving it the well-known chemical formula H_2O. Corresponding to this we also have the molecule of 'heavy water' in which the normal hydrogen atoms are replaced by deuterium atoms.

The same chemical element can therefore have several isotopes. Thus though the number of naturally occurring chemical elements is 92, the number of stable isotopes found in nature exceeds 200. All but 6–7 of the lightest nuclei out of these fall into the category of isotopes made inside stars, and only the remainder are supposed to have been made in the early universe. In the 1950s it had become clear that if we consider the stellar activity during the lifetime of the big-bang universe, say over a duration of 13 billion years, the helium made in stars is only about ten percent of what is actually observed in the universe. The primordial process, on the other hand yields an adequate quantity of helium provided the relationship between density and temperature is properly adjusted.

In 1998 Burbidge and Hoyle questioned the above dichotomy of these processes. Given that the stars do the required job of producing more than 97% of all the isotopes in right quantities, is it not natural to expect them to complete the job by making the remaining 3% also?

Within the framework of big-bang cosmology the answer is no. The time scales are not adequate for stellar activity to have delivered the right amounts of light nuclei. However, in the QSSC, the situation is different for two reasons. Firstly, the MCEs provide locations where the density and temperature are very high for a very short time, and it is possible to have a density–temperature environment in which light nuclei like deuterium, lithium, etc. can be made. Secondly, the longer time scales available in the universe make it possible to make nearly all of the observed helium.

The figure reproduces the earlier diagram showing the variation of the scale factor with time. In the present cycle we have had some 13–14 billion years elapsing since the scale factor was at its lowest value. But preceding that we

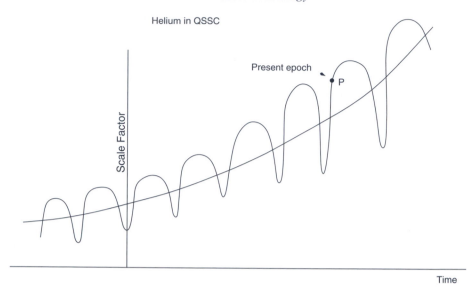

Fig. 15.9. This diagram shows the present and some of the immediately preceding cycles of the QSSC. In the present (incomplete cycle) the helium made from hydrogen in stars is no more than 10% of all helium found in the universe today. However, to this must be added helium made in all previous cycles. Thus, stellar activity in the preceding cycle would make helium from hydrogen. When we add to this the helium produced in all other cycles that preceded the last cycle, the total works out to be a finite quantity and comes close to the quantity of helium usually ascribed to a primordial, very hot era in the big-bang universe. Only, in the QSSC, there was no hot era, but an infinite number of cycles of oscillation each like the present one.

had a complete cycle of 50 billion years or so, in which stars were born and evolved and died. Helium made in these stars would also add to the helium made in this cycle. What about all the preceding cycles? Yes, one should include the contribution of helium made in stars of those cycles also. Since we have an infinite number of such cycles in the past, would we not get the absurd answer that the amount of helium made is infinite? Fortunately not. Just as in the resolution of the Olbers paradox (*see* Chapter 6), here too the long-term expansion of the universe comes to our rescue and we find that the effective contributions of previous cycles get less and less as we go into the more and more remote past. The correct result is that the amount of helium produced by all of the cycles in the past is finite and comes very close to the observed value of 24% or so.

There is a possible problem in this resolution of the helium problem. The normal process of stellar evolution tells us that as stars evolve further, they make heavier elements than helium, like carbon, oxygen, neon, silicon, iron, etc. If the helium abundance has increased in the stellar process, wouldn't the abundances of these heavier nuclei also increase? If they did, then we would end up predicting far greater abundances of such nuclei than actually found. Fortunately, here too the QSSC manages to find a way out. Normally stars come with varying masses and their evolution proceeds according to how massive they are. The number of lighter stars predominates over heavier stars in this distribution. The lighter stars do not proceed beyond helium in making chemical nuclei.

In the QSSC, therefore, most stars contribute helium, but only the massive ones go on to make heavier nuclei. The total contribution from all previous cycles to the net abundance of heavy nuclei is therefore limited, and is consistent with what is observed.

What happens to the stars of previous cycles that have burnt out? The QSSC suggests that these may be the dark matter that astronomers are finding. This will be naturally baryonic. However, the QSSC does not have the same problem that big-bang cosmology had in accommodating baryonic matter. The problem arose in big-bang cosmology because deuterium was made in the primordial era in the universe. In the QSSC deuterium is made either in the MCEs or in flare-type phenomena in stars. Here the number of baryons present is not a crucial factor. In short, the QSSC would be quite comfortable even if the dark matter turned out to be largely baryonic.

Likewise, we may also ask, what happened to the radiation of all those stars from previous cycles, that burnt out? Is it around in any form today? The answer to this question leads us to the solution of the problem of the origin of the microwave background, as we shall see next.

The microwave background

Following the example of helium produced by stars, we may now estimate all the radiation produced by stars in the past. This is a calculation similar in principle to that performed by Olbers for determining the brightness of the night sky. Only we make a realistic assessment of how many stars had shone in a complete cycle and then add up for all cycles including the present incomplete cycle. We then assume that most of this radiation, over the long period of its existence, reached a state of equilibrium. This inevitably happens to any radiation that is lying around long enough and has got scattered, absorbed and re-emitted by matter many times. Such a radiation, as explained in Chapter 12 is known as *black-body radiation*. This radiation is specified by a unique temperature that is determined by how much energy is available in the radiation background per unit volume. The transformation of the original radiation to this equilibrium form is called *thermalization*.

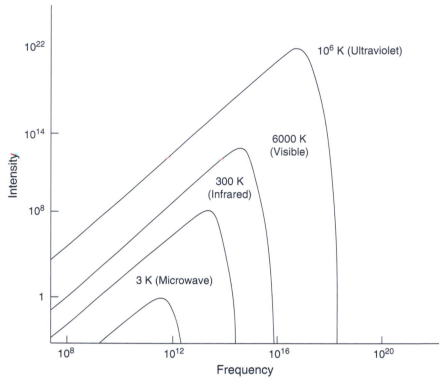

Fig. 15.10. Black-body curves at different temperatures. See Chapter 12 for a detailed discussion of black-body radiation.

The temperature that this relic starlight is estimated to have at present is very close to 2.75 degrees on the absolute scale, *provided it is in black-body state*. We will look at the process that made it possible; but first we need to stress the remarkable closeness of our answer to the observed temperature of the microwave background. At first sight this may appear to be a great co-incidence. Certainly, in the context of the classical big-bang interpretation, this will be so. A closer look will tell us that it is the old calculation of Bondi, Gold and Hoyle in 1955 and the one made by Burbidge in 1958, described in Chapter 12, in a new garb. They had showed that if we assume that all helium were produced in stars then the starlight so generated would on thermalization correspond to the above temperature. This result comes completely from known observational results, and needs no extra assumptions as is the case for the big bang.

So here we have a very plausible interpretation of the microwave back-ground, more plausible than the big-bang interpretation because it gives the correct temperature of the radiation, *provided we have a process for thermalization*.

This process has been under study for the last four decades. In the 1970s, Chandra Wickramasinghe and Fred Hoyle came up with an interesting idea, which has since been developed a good deal further. It is based on the assumed existence of a special kind of interstellar dust, made up of metallic whiskers.

Astrophysicists know that metals are produced and ejected outwards by stars which have advanced to a stage called supernova. Elements, especially iron and carbon are so produced at very high temperatures (ranging from several hundred million to 3–4 billion degrees). When ejected, these are naturally in a vapour form which condenses into a solid as the ejecta move far from the exploded star. The figure below illustrates the scenario.

Laboratory experiments have shown that whenever metallic vapours condense, they do so not as solid balls but as extended very thin whiskers. Typically a whisker may be a millionth of a centimetre in cross-section while being a tenth of a millimetre to a millimetre long. So one expects that the supernova ejecta condense as whiskers which are blown outwards by shock waves generated by

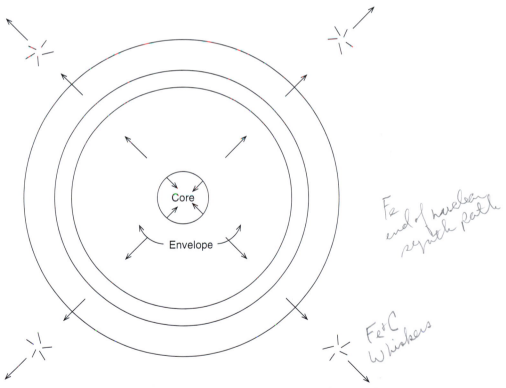

Fig. 15.11. An exploding star (supernova) the gases in whose envelope cool down as they move outwards in the directions of the outward arrows. The gases contain metals which condense into very thin needles or whiskers which are driven outwards into intergalactic space.

the explosions. One can easily visualize that with their high speeds of ejection most whiskers end up outside the parent galaxy and into the intergalactic space. So Hoyle and Wickramasinghe suggested that a regular component of intergalactic dust should be made of whiskers of metals, especially carbon and iron.

Calculations as well as laboratory experiments have also shown that these whiskers absorb radiation mildly at optical wavelengths, are essentially transparent to long radio waves and are most effective absorbers and emitters in millimetre wavelengths. So dust made of this material will have a tendency to absorb starlight and to radiate it out in millimetre waves. Since the microwave background has peak intensity in millimetre waves, these whiskers serve as efficient absorbers of relic starlight from previous cycles and its radiators in microwaves. We expect them to be most efficient in their role when they are most densely packed, which would happen at the epochs of minimum scale factors of the oscillatory cycles.

Is there independent evidence for this kind of dust? We see this in various ways. For example, the radiation from the Crab Pulsar, which is at the centre of an old supernova, shows a gap in its spectrum at millimetre wavelengths. This can be ascribed to absorption by the whisker dust originating in the old Crab Supernova. Likewise there is a similar gap in the spectrum at the centre of our Galaxy, where the supernova activity is believed to be intense. However, unexpected

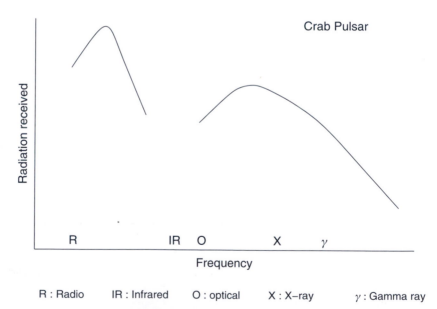

R : Radio   IR : Infrared   O : optical   X : X–ray γ : Gamma ray

Fig. 15.12. The spectrum of radiation from the pulsar situated in the Crab Nebula, the remnant of a supernova explosion. Notice that there is a gap in the spectrum, that is, an absence of radiation. This may be because of the absorption by metallic whiskers emitted by the supernova as explained in the previous figure.

support for this dust came from another observation, namely from the redshift–distance test using supernovae, which we will describe in the next section.

We end this section by describing another feature of the observed microwave background, namely its inhomogeneity on a very small scale. In our present picture, we expect inhomogeneity in the observed background to reflect the inhomogeneity in the galaxy distribution at the time of the last minimum of oscillation. For, it is the galaxies that house stars new or old and the dust will be acting most efficiently at the oscillatory minimum of the scale factor. Any lumpiness of radiation *before thermalization* would continue to show up afterwards also for the last minimum of oscillation. (The relics of oscillations dating further back in time do not show this effect since repeated thermalization would have wiped out any inhomogeneities present in them.) The galactic distribution will show its inhomogeneity in the form of clusters of galaxies. Typically a cluster will show up as a smudged-out disc of more

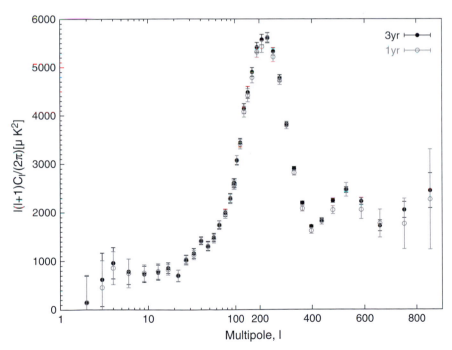

Fig. 15.13. The power spectrum of the cosmic microwave background as reported by the WMAP satellite. The peaks in the spectrum can be interpreted by the QSSC as representing inhomogeneities in the distribution of radiation. For example, the tallest and strongest peak would arise by inhomogeneity on the scale of clusters of galaxies. The smaller peaks to the right likewise arise from weaker inhomogeneities on the scales of groups of galaxies and individual galaxies. In standard cosmology, these peaks are interpreted quite differently. *Figure by courtesy of T. Souradeep, R. Shah and P. Jain.*

intense radiation than its surroundings. The angular size of the disc can be computed using the QSSC model; and lo and behold, a major inhomogeneity seen in the radiation matches exactly with this effect.

At this stage a big-bang supporter may object: Are we not invoking a rather special type of dust to explain the effect? It does not look very plausible to imagine that space is filled with iron whiskers. This objection can be countered in two ways. Firstly, the density of dust required is extremely small, a few thousandth parts of metals normally found in the universe. And as we will further show, it is consistent with the results from the redshift–distance test of supernovae. Secondly, the origin and physical properties of the dust are astrophysically explained and experimentally observed. So on the credibility scale the whiskers should certainly have a higher rating than non-baryonic dark matter. Recall that NBDM is assumed to exist in far greater proportion than ordinary baryonic matter that we know, although it has never been found and its properties, if it exists, are completely unknown.

Back to Type Ia supernovae

We now return to the test described earlier, in Chapter 14, the use of Type Ia supernovae to relate redshift to distance in order to decide how fast the universe is expanding and whether the expansion is accelerating or slowing down. In the big-bang models, the matching of data with theory required many alterations to be made in the latter. It required adopting a cosmological constant to begin with, then a variation of this 'constant' with epoch, the rather finely tuned adjustment of proportions of baryonic matter, NBDM and the present strength of the cosmological constant. Since the big-bang models demand a positive cosmological constant, the expectation of the big-bang supporter was that the QSSC with a negative cosmological constant was bound to fail this cosmological test.

Quite the contrary! There was another feature of the QSSC that came to its aid. As we saw in the discussion of the cosmic microwave background, the intergalactic space has dust in the form of whiskers. It will absorb part of the radiation from distant objects. This will make distant supernovae look fainter by comparison than if the dust were absent. Effectively therefore, the presence of dust has the same qualitative consequence as the positive cosmological constant had for the big-bang models.

The next figure shows how the QSSC prediction fits the data on a specially selected 'Gold Sample' of supernovae. The fit is technically not bad considering the amount of whisker dust that is needed for thermalization of the cosmic microwave background.

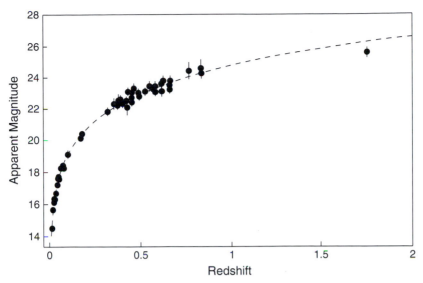

Fig. 15.14. The redshift–magnitude plot of Type Ia supernovae, described in Chapter 14, is reproduced here along with a curve (dotted line) drawn according to the QSSC prediction based on absorption produced by intergalactic whisker dust.

In fact one can do the data-fitting exercise differently. We do not put in any specified dust density. Rather we choose that density which gives the best fit to the data. This best-fit value of the density turns out to be fully compatible with the thermalization process needed for producing the microwave background. The supernova test and the microwave background are totally independent tests and the fact that the QSSC passes both with the whisker dust density in the same range of consistency increases our confidence in the overall picture.

QSSC versus the standard model

We end this chapter with a brief comparison of the QSSC and the standard big-bang cosmology (SBBC in brief).

The SBBC explains the origin of light nuclei and the cosmic microwave background in a framework that requires many speculative assumptions about what went on in the very early universe. In the QSSC, the untested theory is about the creation process, although the *C*-field is just one assumption made. In the SBBC, the theoretical assumptions are many, e.g., inflation, NBDM, GUTs, quantum gravity era, dark energy etc. The QSSC also explains the microwave background and in this process establishes a link between the astrophysics of starlight, nucleosynthesis and intergalactic dust. All elements of this explanation are testable and observable.

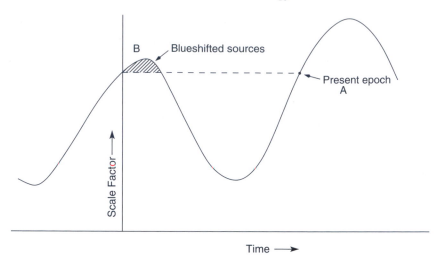

Fig. 15.15. The QSSC expansion curve is shown with two consecutive cycles: the current cycle and the previous one. A is the present epoch on the current cycle, and B is the epoch when the previous cycle had the maximum scale size. Notice that a galaxy at B when seen by an observer at A will show a blueshift in its spectrum. In fact all epochs of the previous cycle shown in the shaded region will have galaxies which will be seen blueshifted at epoch A. This is because, at these epochs the scale size of the universe was larger than at present.

The QSSC predicts very faint objects with blueshifts. In the above figure showing the expansion of the quasi-steady-state universe with time we see that there are sources of light in the previous cycle when the scale factor was greater than the present scale factor. Light coming from any of these sources will be blueshifted. That is, the wavelengths of light of specific elements will appear to be *shorter* than their laboratory values. Being in the previous cycle, these sources are far away and as estimated now, they will be barely observable with the best telescopes of today. The shifts expected are small, around 10%.

Since blueshifts are not predicted by the SBBC, this will be a good test for deciding between the two models.

Another good test concerns searches for very old stars. In the SBBC, the universe is around 13 billion years old, so we do not expect to find any object, a star or a galaxy, older than this. Also since the galaxy formation process was over several billion years ago, we do not expect to see very young galaxies. The QSSC on the other hand is ageless and can in principle have objects very young and very old. In particular, stars of very low mass, say half the solar mass, born in the previous cycle would be around today and may have become red giants. The findings of such low mass giants would mean that they are 40–50 billion years old and such stars are not possible in SBBC. Likewise the QSSC will allow very old white dwarf stars, older than 20 billion years that cannot exist in the SBBC.

Finally, a deciding factor may be the nature of dark matter. As we have seen, finding dark matter in excess of density 4% of the closure density in a baryonic form will pose problems for the SBBC. If non-baryonic dark matter is found in abundance, the SBBC can take credit for having predicted it, although its presence does not rule out the QSSC.

There are other aspects of a more technical nature which we have not gone into. But what we have already mentioned is enough to show that the QSSC provides a testable and viable alternative to the SBBC.

However, the reader should be aware that many of these tests are being attempted, but that most of the observers who make them know that they will get more favourable treatment from their colleagues, editors, funding agencies and others who assign telescope time if they concentrate on tests confirming the big bang. For example, no time has ever been assigned on the Hubble Space Telescope to observers who are thought to favour the QSSC, and *everyone* who designs, plans, builds, and observes for the microwave background believes from the beginning that it originated in the big bang. So much for unbiased observers.

We could have ended the book here: but this would have been unfair to a whole class of data of cosmological significance that have been steadily accumulating since 1970, but which have been consistently ignored as they apparently cannot be explained within the standard paradigm. Even the QSSC has no ready-made solution for these findings.

We will outline these data in the next chapter.

16

Unfaced challenges in cosmology

Do we really understand redshifts?

All along we have assumed that observations of cosmological redshifts imply that the universe is expanding. This was the interpretation given in the early 1930s to the Hubble law, and indeed at the time the observations did seem to fit the hypothesis of the expanding universe very well. Further extensions of Hubble's original results after he died in 1953 to fainter (and therefore inferred to be more distant) galaxies, by later workers like Milton Humason (his original collaborator), Nicholas Mayall and Allan Sandage, also lent support to the expanding universe picture. Indeed, the whole of the cosmological scheme described in this book rests on this premise. Not only the standard big-bang model but also the alternative, the quasi-steady-state model assume that the redshift of an extragalactic object arises from the expansion of the universe.

There is, however, a fly in the ointment! Indeed, it is a fly that we believe cannot be ignored. There are observations of the extragalactic universe which may cast doubt on the above holy cow of the expanding universe. These observations began to surface in the 1970s, and despite the early hope of most cosmologists that with better observing techniques, they would go away, they have stayed. More disturbingly, they have proliferated. Originally found with conventional optical techniques, these observations are today seen in objects identified with some sources of radio, X-ray and gamma ray emissions.

Collectively, the phenomena are put in one box labelled 'anomalous redshifts'. We devote this chapter to opening this Pandora's box. Most cosmologists would like to keep the box locked up. . .for the issues that these phenomena raise might ultimately force them to seriously modify the basic paradigm.

While we believe that the expanding universe hypothesis is correct, we also believe that there are some objects with anomalous redshifts. Thus we consider that the tactic of ignoring these phenomena is unwise. There have been

Fig. 16.1. This pair of galaxies shows tidal interaction between them, with the smaller one pulling material away from the bigger companion, M 51. Credit: *NASA, Hubble Heritage.*

instances in the past where new observations did not fit within the accepted framework and ultimately forced new ideas upon scientists. These new ideas then led to a healthy growth in the subject. The quantum theory of matter and radiation began in this way, with experiments on the microscopic level yielding results that could not be explained by the established theories of Newton and Maxwell. Radical departure from well-accepted concepts had to be admitted into physics to make way for these apparently anomalous phenomena.

In astronomy and especially cosmology, we work at the frontiers of our current understanding. Observations take us to uncharted territories. While apparently anomalous phenomena might arise from observational artifacts or errors of interpretation, one needs extra caution while probing anomalies that remain and do not go away. We feel that there are many such phenomena that need further careful studies. Ignoring these would be like throwing the baby away with the bath water.

We come now to specific cases.

Close pairs and groups of galaxies

If two galaxies appear to be close together in the sky and if they have about the same redshift, it is reasonable to suppose that they are physically close to each other. If they are on different orbits and are coming together they will ultimately collide – they are merging. But in general we don't know what their orbits are. In one case, that of the Milky Way and Andromeda we do, and they are approaching each other at about 300 kilometres a second. Thus they will ultimately collide.

But in general how can we tell whether the galaxies are going to merge, or move apart? If they are coming together, tidal tails caused by gravity will extend at great length outside them. They are faint but they can be seen in some comparatively nearby pairs. However if they have been born together they may be either in orbit about each other or they may simply be moving apart. If they are moving apart, there will be no tidal tails, but there may be a great deal of nuclear activity, and evidence for explosions in the central parts at one or both galaxies.

It was Ambartsumian who proposed in the 1960s on observational grounds that groups of galaxies were moving apart and that in essence, one galaxy was born in the centre of another. This is a very unpopular idea but it fits in well with the QSSC. We predict that matter and energy are being created and ejected from the centres of active galaxies.

None of us really knows how galaxies are made. However, in the conventional big bang protogalaxies are formed, and galaxies as we know them today are built up by mergers of smaller galaxies.

For the faintest galaxies, those furthest away, we cannot detect tidal tails even if they are there. But because this scenario prevails, literally hundreds of papers are written based on the merger hypothesis and almost none on the ejection (explosive) process. All of the theory is devoted to mergers. Almost certainly there are events of both kinds taking place in the universe. But since big-bang enthusiasts don't know how to make galaxies without mergers, only one aspect of the problem is being studied.

There are two nearby galaxies, Arp 220 (*see* Figure 16.6) and NGC 5128 which are both very active, are ejecting QSOs and other systems, are radio sources and have jets, showing that they are exploding at present. However, they are frequently claimed to be two galaxies which have collided.

We turn now to some well-known individual groups beginning with Stephan's quintet. This is a group of five galaxies discovered in 1788 by E. M. Stephan in France. Four of the galaxies have redshifts nearly the same, but in 1961 Margaret Burbidge found that the fifth galaxy has a very much smaller redshift.

From this we can draw one of three possible conclusions. First we can say that the odd one out has really no dynamical connection with the rest; that it just happens to be passing through the group of the other four. This is like four

Fig. 16.2. This image describes the five galaxies forming Stephan's quintet discussed in the text. Credit: *Chip Arp*.

people standing chatting in a park while a fifth person happens to pass close to them while jogging. If caught on camera just at this instant, the photograph may convey the erroneous impression that the fifth one is part of the group. The second conclusion is that the odd one out is located at its correct distance according to Hubble's law, but happens to be projected close to the rest of the group. This is like the occultation of Venus by the Moon...at the time of occultation, Venus appears very close to the Moon, but is in fact much further away.

Both of these explanations are possible, although, as we can see they involve somewhat contrived situations that are not likely to happen frequently. Thus the jogger in the park being photographed with a stationary group or Venus being occulted by the Moon are events that have a low probability of occurring if looked at at random. So if we consider such contrived scenarios as unlikely, what is our third and remaining conclusion?

This conclusion is that the odd one out in Stephan's quintet does not have its redshift determined by Hubble's law. Then it is a member of the group, but has an anomalous redshift, i.e., a redshift not following Hubble's law. That this case is not unique has been shown by the observation of other such examples, including the quintet VV 172 in the catalogue of peculiar galaxies prepared by Vorontsov–Velyaminov in the USSR some fifty years ago. The same situation exists with regard to this quintet as with Stephan's quintet. In the subsequent years a catalogue of about 100 compact groups of galaxies each involving 4 or 5 galaxies was made, and in about 40% of the cases, one galaxy has a redshift much larger, or smaller, than all of the rest. This seems to us to indicate that in many cases there are anomalous redshifts, or that the galaxies are literally exploding away from the group where they were born. But this is not the Establishment (popular) view. In *all* cases it is generally argued that the galaxy with a discrepant redshift is either a background or foreground object and the apparent association is due to accidental projection. Of course, the probability that this is the explanation is very small.

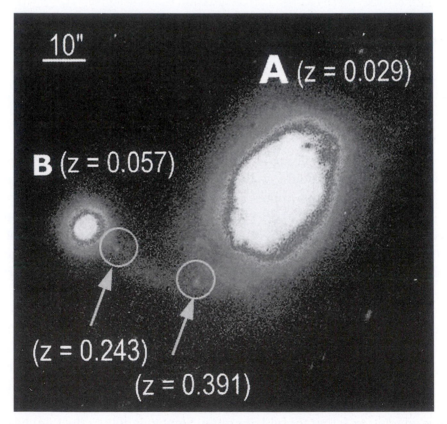

Fig. 16.3. The galaxy with the catalogue number NGC 7603 and its companion have redshifts very different from what one would expect from Hubble's law. Credit: *Lopez-Corredoira and Gutierrez, whose work is cited in the text.*

Another remarkable situation is found near the bright spiral galaxy NGC 7603. This was originally found by H. C. (Chip) Arp and a picture is shown here (Figure 16.3).

This galaxy is clearly connected to another smaller galaxy by a bridge of gas and stars, but the smaller galaxy has about twice the redshift of NGC 7603. Is this difference due to velocity (very unlikely) or something else? In the bridge joining the two galaxies there are some very small compact objects, and two young Spanish astronomers M. Lopez Corredoira and C. M. Gutierrez have recently shown that they are small galaxies with redshifts about five times larger than NGC 7603.

Quasars and galaxies

Are there yet more accidents? We turn now to the redshifts of the quasi-stellar objects (QSOs) or quasars. When they were first discovered by radio and optical astronomers they were found to look like stars but to have very large redshifts.

It was soon found that they did not follow the Hubble law – there was no correlation between their brightness and their redshift. Nevertheless, if they were

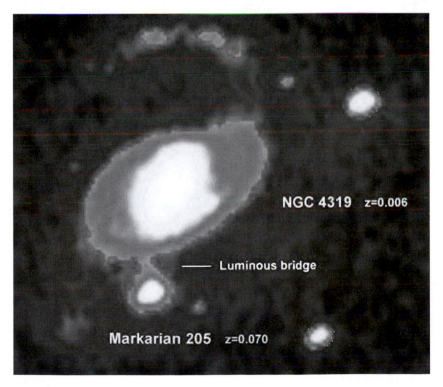

Fig. 16.4. This pair of galaxies, NGC 4319 and Markarian 205 show discrepant redshifts as described in the text. Credit: *Chip Arp*.

very far away as would be the case if the redshift was a measure of distance they could be used for cosmology, and this is what most astronomers wanted to use them for. Thus it was argued that they were really the active nuclei of galaxies and were so far away that the outer parts of the galaxies were too faint to see. In other words, there is a continuity between nearby active galaxies and quasars – each of which must have a host galaxy. This is still what most people want to believe today. However, much observational evidence suggests otherwise.

In 1970 a very close pair, an active galaxy and a quasar, NGC 4319 and Markarian 205, were found to be very close together; Chip Arp and Jack Sulentic showed that they are joined by a luminous bridge. Also, Mk 205 has more than ten times the redshift of NGC 4319. They are shown in Figure 16.4.

This is one of the few anomalous cases which have been photographed using the Hubble telescope and where more than one set of observers have accepted that the bridge is real.

This means that the difference between the redshifts of Mk 205 and NGC 4319 must be due to an effect so far not understood. Hubble's law has failed here as it has for NGC 7603 and its companion.

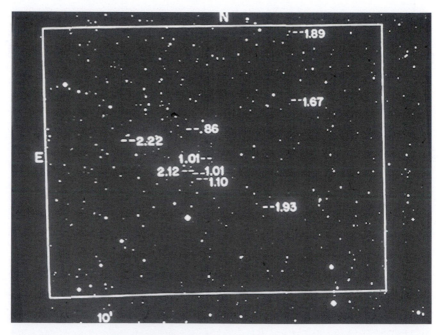

Fig. 16.5. A photograph showing the unusual concentration of quasars in the immediate vicinity of the galaxy NGC 3810. The galaxy has a very low redshift, of the order of 0.003, whereas, the quasars have considerably higher redshifts as shown in the figure by numbers. Photo Credit: *Chip Arp.*

ULIRGs

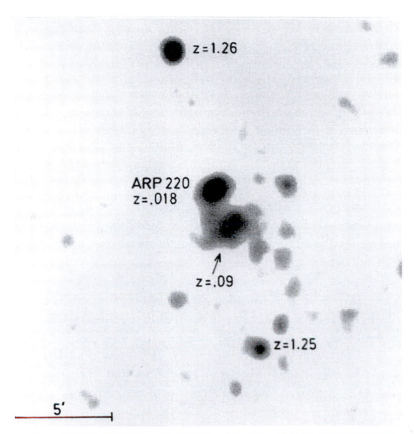

2000 - INNER REGION OF ULIRG ARP 220. X-RAY QUASARS EX-
ACTLY ALIGNED WITH ALMOST IDENTICAL SPECTRA. TRAIL OF
X-RAY SPOTS LEADING DOWN TO z=1.25 QSO. GROUP OF Z=.09
GALAXIES CONNECTED BY X-RAYS AND LOW z=.018 HYDROGEN.

Fig. 16.6. The ultra-large infrared galaxy Arp 220 is shown with two X-ray quasars of
almost equal redshift aligned across it. The redshifts of the objects are shown in the
image. Credit: *Chip Arp.*

If quasars do lie at cosmological distances they are very powerful, since they
are emitting more energy than a whole galaxy of 10^{10} stars from a size that is
not much greater than the Solar System. Is this possible? The conventional
wisdom is that this takes place because the energy is released by matter falling
into a massive black hole. The brighter the object, the more massive must be
the black hole, and the more the stars fall into it. Pushing Einstein's theory to
the limit, and extrapolating from very nearby active galaxies in which much

lower mass black holes are found it is claimed that the problem can be solved. But it is what is called a bootstrap argument. For a cosmologically distant quasar we cannot detect, or measure, the mass of a black hole, nor see the galaxy that needs to be present. Underlying this whole 'theory' is the old adage

'We've tried so hard, we've tried so long
We must be right, we can't be wrong.'

But if quasars are much closer than their redshifts suggest, their luminosities are much lower, and the energetic requirements are much less.

Quasars are few in number compared with galaxies. Thus to find these clustered around near to active galaxies would not be expected. However, in the last few years, Arp together with others in our group has found many cases of this kind.

In each case a scientific paper is prepared, and after a lot of criticism from referees is published. Those who read it are astonished and/or disbelieving, and consider that it must be an accident. But above all they think that these are isolated instances and ask for more. A year goes by and another case appears and we go through the same discussion. Physicists don't like statistical arguments since there is always considerable uncertainty attached to them. But in these cases the chance that each of these configurations is accidental is 1 in 10 000 or less!

Those few of us who are serious gamblers don't really understand this. If you went to a racetrack and only bet on the horses that were quoted to win at odds of 100 to 1 or greater, you would be considered to be a great fool. But in physics you would be considered to be a wise man.

The trouble is that for the conventional cosmologist to accept even a single case would impair the credibility of Hubble's law. The discovery of the phenomenon of gravitational lensing in the 1980s seemed to provide a tool for the conventional astronomers to answer Arp's criticism of Hubble's law. Gravitational lensing implies that light rays are bent by the strong gravitational attraction of a galaxy. This can refocus the rays from a source of light so as to make it appear brighter than it is. So the quasars seen around the galaxy may look brighter than they actually are. Since fainter quasars are more thickly populated, the chance of finding them around a (lensing) galaxy is enhanced.

This argument seemed at first to raise the poor odds estimated earlier and improve the credibility of Hubble's law. However, it became clearer that this prescription did not work so well for all quasar–galaxy associations. It worked for only very few which had quasars lying very close to the galaxy.

Chip Arp has devised the following parable: A cosmologist and a forest ranger went into a forest together. Towards nightfall, the ranger spotted what

Fig. 16.7. Chip Arp, who has played a pioneering role in discovering systems with discrepant redshifts. Credit: *Chip Arp*.

he believed to be a tiger. 'Look, there is a tiger in the vicinity. Follow me. We should climb a tree quickly.' And he did so. The cosmologist, however, said: 'This tiger is not a local tiger. My calculation shows that he is quite distant, and happens to be projected against these trees.' So he remained on the ground until the tiger came and ate him.

Alignments

In radio astronomy, we have typical double sources with a central galaxy or quasar with two blobs on opposite sides showing major regions of radio emission. Activity in the central regions has ejected plasma, that is electrically charged particles, in opposite directions. The jets of particles impinge on intergalactic medium and slow down. The space contains magnetic fields also and the particles radiate what is known as 'synchrotron' radiation which is mainly in radio wavelengths. The essential point for our purposes here is that the alignment of the two radio-emitting regions on both sides of the central region implies that material has been ejected in opposite directions. The alignment of the blobs across the central regions is taken as indicative of ejection by the central galaxy.

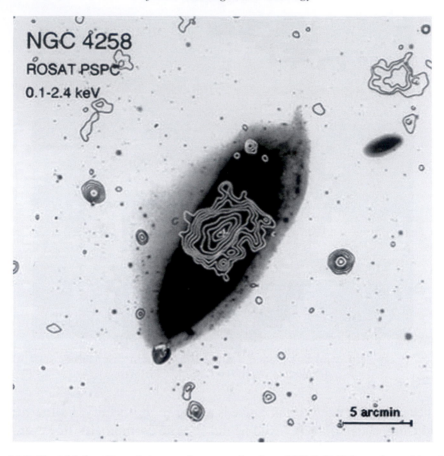

Fig. 16.8. Two blobs aligned across the central galaxy NGC 4258 have been identified with two quasars of redshifts 0.40 and 0.65, far more than the redshift 0.001 55 of the galaxy. Credit: *Chip Arp*.

Consider now the following case illustrated in the figure. Here we have two quasars aligned across a galaxy. Can we likewise interpret the configuration as an example of ejection of two quasars in opposite directions by the galaxy? No conventional cosmologist would accept this interpretation, because, the red-shifts of the two quasars are quite different from that of the galaxy. According to Hubble's law, the three objects are in different parts of space and just happen to be projected so that the alignment is achieved. If this happened at random, we would not expect any alignment. Indeed the probability of this happening by chance is very small...unacceptably small for our statistician testing the explanation.

We next see *two* triplets of quasars with each triplet exactly aligned, but no galaxy. The alignment is perfect and well within the small error-bars of

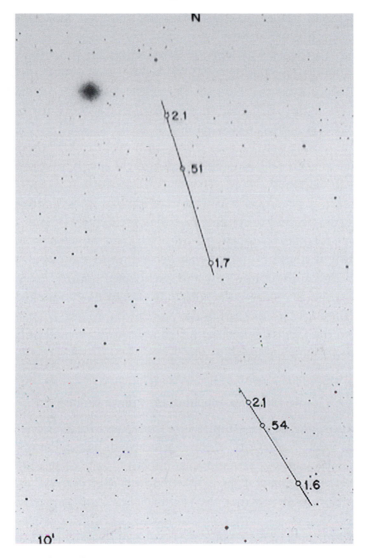

Fig. 16.9. The triplets of quasars discovered by Chip Arp and Cyril Hazard in 1980. See the text for details. Credit: *Chip Arp*.

observation which was very accurate. The quasars in each triplet have different redshifts. But additionally the redshifts of quasars in the two triplets seem to match reasonably well. Since the triplets were found reasonably close to each other on the same photographic plate, it is not unlikely that the entire configuration is part of one system. Perhaps the two central quasars were ejected first and then they in turn ejected pairs in opposite directions.

However, all this ejection business becomes untenable if one is looking at the redshifts of the quasars as arising from the expansion of the universe. We are then talking of projection effects with the observed configuration arising from our perspective with a very very small probability.

Something has to give way when the Establishment is bombarded by such anomalous findings. One fine day in 1984, Arp received a rejection letter for his proposal to observe on the 200-inch Palomar telescope, on which as a staff member he should have had guaranteed access. The letter conveyed the regrets of the observing committee, which felt that as his findings did not make sense, he should not be allowed time for more observations of this kind on the telescope. He appealed against the decision, but it was upheld and in frustration he left his job to move to Germany.

Of course, in a peer review system one aims to pick good and exciting observing proposals and reject all those that are cranky or not well thought out. But when it comes to rejecting a proposal from an experienced observer, simply because it does not make sense according to the popular belief, one needs to be worried about how new and unexpected findings are ever going to be made. Astronomy, indeed, the whole of science has flourished because of unexpected findings that did not fit the current paradigm. In essence, all this boils down to the fact that our best instruments will be used only to observe *what is expected to be observed.*

Can we try other explanations known to physics?

While the whole gamut of examples like those described here is hard to understand within the framework of the expanding universe, the question arises as to whether we can invoke any other known source of redshift to explain these observed anomalies? That is, if we admit that there is an anomalous redshift that cannot be explained by Hubble's law, can we discover the cause of it?

One such known source of redshift is gravity. Imagine that two observers are in communication with each other. Observer A is situated far away from a massive compact object on which observer B is located. Then Einstein's theory of relativity tells us that in comparison with A's watch the watch of B will run slower. This is because the force of gravity is much more powerful on B than on A. The result is that the light emitted by B will appear redshifted to A. Conversely, B will find the light emitted by A to be *blueshifted.*

Imagine that we are observing a quasar and a galaxy which are real physical neighbours. The quasar is a compact object with a strong gravity, whereas the galaxy may not have such a strong gravitational force acting around it. So the quasar light will have an *extra* redshift arising from its strong gravity. Could this be the anomalous redshift we are observing?

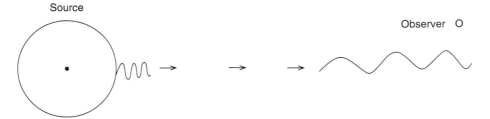

Fig. 16.10. The light wave emitted from the surface of a compact and very dense star is shown as it travels to a distant observer O. The wavelength of the light wave increases by the time it reaches O. This phenomenon is known as gravitational redshift. In this case the redshift relates to the light from the surface of the star.

At first sight this may be a promising line of attack. However, it does not explain what is observed. For, very careful theoretical calculations by the mathematician astronomer Hermann Bondi (one of the authors of the steady-state theory) tell us that any astronomical body held in equilibrium under forces of gravity and outward pressures of gas or radiation, will have a maximum redshift from its surface, no more than 0.7 or so. The anomalies that need to be explained are much larger, certainly as high as 5–6 in the redshift.

Fred Hoyle and Willy Fowler once proposed that for a compact massive object, the gravitational redshift of light may be limited to what Bondi found,

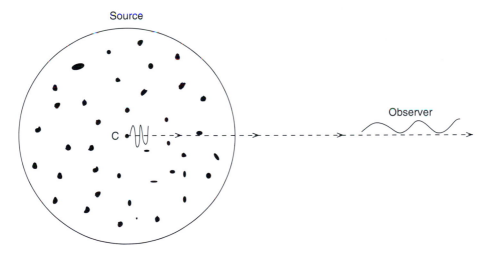

Fig. 16.11. In contrast to the preceding figure, here we have light emerging not from the surface of the star but from its interior. In this case the gravitational redshift can be higher: but the star will have to be rather unusual in its make up, to allow light from the interior to escape without significant absorption. (In an ordinary star, the matter inside the star absorbs the light emerging from its interior.)

Fig. 16.12. The quasar ejected by the galaxy is moving away from the observer, and will be seen redshifted. However, some quasars could be ejected towards the observer and should be blueshifted. Why have no such cases been reported?

but it could be much higher if the light left from its interior. P. K. Das from India in fact showed with specific models that anomalies of redshifts of the order of 2–3 could be explained in this way. However, how can light emerge from the interior of a compact dense object? Would it not be stopped on the way out? Scenarios allowing light from the interior to emerge have been suggested, but they seem very contrived.

What about the Doppler effect described in Chapter 6? Indeed this was the first to be suggested. If a galaxy has ejected another or a quasar, then the ejected component may be moving rapidly relative to the galaxy and would therefore show redshift larger than the redshift of the galaxy. But there is a problem here too. A random observer situated in space may see the ejecta coming in his direction as well as in the direction away from him. In the former case the ejecta should show a *blueshift* relative to the galaxy. So far hardly any case of such a blueshift has been seen.

So the gravitational and the Doppler options both seem to be inadequate in explaining anomalous redshifts. If we assume that the Hubble law applies to most galaxies but fails for companion galaxies or quasars, then the failure is nearly always in the same direction: the companion objects seem to have larger redshifts. The Doppler option fails on this count. The gravitational option is not able to deliver anomalies in redshift as large as are often observed in quasars.

Can we then look for another option that has not yet been put to the test?

The variable mass hypothesis

In 1964, Fred Hoyle and one of us (JVN) proposed a new way of looking at the phenomenon of gravitation. This approach was inspired by the ideas of Ernst Mach, a philosopher–scientist of the nineteenth century. The so-called 'Mach Number' used to describe the speed of a supersonic aircraft is named after Mach. The Mach number tells us the ratio of the speed of the aircraft to the speed of sound. Mach had been very much interested in the basic concepts of dynamics, in particular the laws of motion first stated by Isaac Newton in the seventeenth century. Mach found the Newtonian discussion of these laws inadequate, especially concerning the origin of inertia.

We know that if we find a football lying in our way, we can dislodge it and send it away by a gentle kick. This would not be adequate if we had a football-sized object made of solid iron. Much greater force would be needed to move it. What is the property that makes this distinction between a football and a ball of iron? The property is known as 'inertia', or the tendency of the body to resist change in its state of rest or motion. Newton argued that this is an intrinsic property of the object, and can be measured by its mass. The larger the mass of a body, the greater its inertia, and the greater will be the force needed to change its state of rest or motion. A body of mass 100 kilograms will have much greater inertia than a body of mass one kilogram. Newton went on to describe how one relates the mass to applied forces and resulting motions.

Where Mach disagreed with Newton was in the latter's assumption that inertia is the intrinsic property of matter. Mach argued that the property of inertia owes its origin to the large-scale structure of the universe. It is the background of distant stars or galaxies that provides the framework in which to measure local motions. Without a background we cannot detect or quantify how fast a local body is moving. Mach's principle, as this idea is often known as, therefore prompts a deep look into the origin of inertia.

Although Mach gave very persuasive arguments and observational results in favour of his principle, the details of which we cannot go into here, he did not give a quantitative theory as to how to incorporate his ideas in a mathematical theory of motion and gravitation. His ideas have had considerable influence on many distinguished physicists, including Einstein, who looked upon Mach as a teacher. At one time Einstein hoped that his general theory of relativity would turn out to incorporate Mach's ideas. This did not happen, although, following him other physicists attempted to incorporate Mach's principle into a theory of gravity.

The Hoyle–Narlikar theory of gravity (HN-theory in brief) was one such attempt and it showed that starting with a basic formula that relates the mass of any particle of matter to all other particles of matter in the universe, one can arrive at a theory of gravity that looks very much like the general theory of relativity. . .except for some situations wherein differences arise. This tells us why Einstein was tantalizingly close to incorporating Mach's principle into general relativity, but did not actually reach that goal.

The exceptional circumstances where new effects can be seen in the HN-theory that distinguish it from general relativity, were first discussed by one of us (JVN) in 1977. The circumstances relate to the creation of new matter. In

Fig. 16.13. Ernst Mach (1838–1916).
Department of Physics, Texas A&M University – Commerce.

the steady-state theory and the quasi-steady-state cosmology, creation of new matter plays a crucial role in determining the large-scale properties of the universe. A subtle modification of the basic framework of the HN-theory leads to the following important consequence.

Whenever new matter is created, it starts with zero inertia, since at the time of creation it has not had time to interact with the whole of the universe. Since the interaction that conveys and contributes to inertia of new matter travels with the speed of light, its effect on the mass of the new matter gradually builds up. As the matter ages, it comes into contact with more and more of the universe and so its inertia builds up. In short, we have new matter having mass that does not stay constant but increases with time. For this reason this idea is referred to as the *variable mass hypothesis*.

How does it lead to anomalous redshifts? The answer to this question can be understood by imagining an atom of hydrogen, in which the electron and proton both have half the masses that they are known to have in a typical hydrogen atom in the laboratory. Calculations of atomic theory applied to this atom will tell us that the wavelengths of light emitted by this atom in

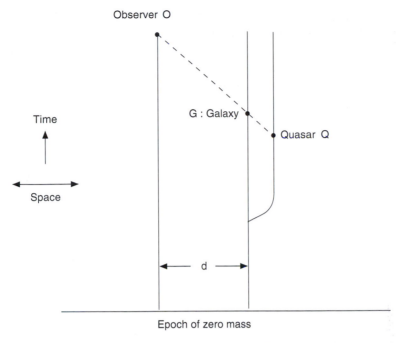

Fig. 16.14. The variable mass hypothesis (VMH) uses the Machian idea that the inertia of a particle is zero at the time of its creation and increases as it ages, since it begins to receive contributions to its mass from the rest of the particles in the universe. The contributions travel with the speed of light and those from the more distant parts of the universe arrive later. Thus the mass of a created particle starts growing from the moment of its creation. In the figure the worldlines of two galaxies are shown in the spacetime diagram. The observer is on galaxy *O* while *G* is another galaxy at a distance *d* from *O*. Assume that both these *galaxies* were born at the same epoch so that at any epoch the particle masses in both galaxies will be the same. However, when the observer on *O* sees *G*, he sees it as it was in a past epoch. Because, light from *G* will take time d/c to travel to *O*, *c* being the speed of light. So *O* sees the particle masses in *G* to be smaller than in his laboratory. The proof for this result will be the wavelength of light he receives from *G*. For, atomic physics tells us that the wavelength of light emitted depends on the mass of the particle emitting it: the smaller the mass the longer the radiation. So to the observer on *O*, the light from *G* appears redshifted. Notice that this redshift is not arising from the expansion of the universe but because of particle masses increasing with age. Next in the figure we also see a quasar *Q* created and ejected by *G*, at a relatively recent epoch. So particle masses in *Q* will be much smaller than those in *G*. Hence if the observer examines and compares the spectra of both *Q* and *G*, he would find that the redshift of *Q* is higher than that of *G*, even though they are close neighbours.

its spectrum will be *twice* the wavelengths of the same lines in a normal laboratory atom. This is an example of the general rule that the younger the piece of matter, the longer the wavelengths of the characteristic lines in its spectrum.

The presumption here is that the atom studied in the laboratory is old and has a larger mass than a newly created atom. So, if we imagine an old galaxy creating and ejecting a quasar, the spectral lines of the quasar will be longer than the spectral lines of the galaxy. So whatever the redshift of the galaxy, the quasar will show a higher redshift. The extra redshift that the quasar has is the anomalous component that we have been talking about. And, from the above argument, the younger the quasar, the larger the anomalous redshift. This is a consequence of the variable mass hypothesis.

This hypothesis was applied by Prashanta Das and JVN in 1980, to a number of quasar–galaxy associations as well as to galaxy–galaxy associations. The observed features tallied very well with the theoretical descriptions. Later work by Chip Arp showed that when one sees a number of quasars ejected by the same galaxy, one finds that the further quasar has the lower anomalous redshift. This happens because as the quasar ages it moves further away from the parent galaxy.

In this picture, an ejected quasar may move further and further away from the parent galaxy if the force of its ejection is greater than a critical limit. Such quasars may be identified with those that seem isolated with no galaxy nearby to relate to. If the ejection force is less than the critical limit, the quasar comes to a halt and returns close to the galaxy, finally going around it in smaller and smaller orbits. As it ages, its mass grows and it evolves into a galaxy. Thus a companion galaxy close to a main galaxy like NGC 7603, say, can be seen as an 'older version' of the typical quasar–galaxy association.

The variable mass hypothesis therefore seems to provide a basic explanation of what has been described so far. However, the story does not end here! There are more weird observations relating to redshifts that pose challenges even to this hypothesis. We will come to these next.

Periodic redshifts

Let us look at a series of numbers: 2, 7, 12, 17, 22, 27,... What special property does this series of numbers have? We find that each number is 5 more than the previous one in the series. If a certain event were to happen on the dates of the month as given by these numbers, we would say that the event recurs with a period of 5 days. In general, where we have a series with numbers repeating after a fixed interval we may call the series an *arithmetical* series.

Next look at another series: 2, 6, 18, 54, 162,... What is special about the numbers here? After some inspection we discover that each number is 3 times the previous one in the series. Here also we may talk of a period 3, but in the sense of multiplication rather than addition. Such a series is called a *geometric* series.

The purpose in introducing these two special series is to highlight the fact that the numbers in each series are specially selected. A series with numbers chosen at random will not have a period either of addition or of multiplication. If a natural event shows a period then we should suspect that there lies some deep reason to be unravelled. In Chapter 3, the finding of a period in the arrival dates of comets by Halley led him to the discovery that it was one and the same comet coming close to the Sun every 76 years. From this finding he argued that comets are subject to the same gravitational force of the Sun as the planets.

In short, any pattern in numbers measuring some natural phenomenon holds a clue to the phenomenon and should be studied carefully in order to uncover the scientific law underlying the phenomenon. The discovery of the pattern of wavelengths of emission lines in the spectrum of the hydrogen atom led Niels Bohr to the first clue to how the rules of quantum mechanics control a microscopic system like the hydrogen atom.

It is against this background that cosmologists are today being challenged to probe, in order to prove or disprove, the claim that the redshifts of many chosen systems do not just have any values, but seem to follow some kind of periodicity, either of addition or multiplication.

The story began for redshifts of galaxies in the year 1976, when Bill Tifft at the Steward Observatory, Arizona analyzed the redshifts of galaxies in the Coma cluster.

In a cluster, galaxies are moving at random under one another's force of gravitation. So if we assume that in addition to the expansion of the universe (the Hubble law), these motions contribute to additional redshifts of these galaxies, then we expect the differences in redshifts to be randomly distributed. In short we should not expect to see any pattern in the differences of redshifts of all galaxies. Contrary to this expectation, Tifft found a strong signal that the redshift differences were distributed with a small period. If we multiply the redshift differences by the speed of light we get a period in the form of a velocity of 72 kilometres per second. This velocity unit for redshift is sometimes used in astronomy, especially where small redshifts are involved.

Tifft was certainly surprised by this unexpected finding and very carefully checked the data and his method of analysis. The result did not change. Not only that, when he later applied it to other classes of objects like pairs of galaxies, satellite galaxies and central galaxies of small groups, the period was quite clearly there. Later in 1985, Chip Arp and Jack Sulentic used very accurate measurements of redshifts of the 21 cm line of hydrogen (*see* Chapter 14) in galaxies to confirm the Tifft result.

Fig. 16.15. The Coma cluster of galaxies. Credit: *Atlas Image courtesy of 2MASS/ UMass/IPAC-Caltech/NASA/NSF.*

How has the cosmological community been reacting to these findings? The first response was to doubt the credibility of Tifft's data and then of his analysis. However, very rigorous examination of the data by Bill Napier who used sophisticated statistical techniques revealed that the periodicity was really there. In fact Bill Napier started off as a sceptic of Tifft's analysis, only to end up confirming the validity of the results. In 1996 Guthrie and Napier confirmed periodicity but found a strong signal for a period of 37.6 kilometres per second (using velocity units as explained before). This figure is nearly half of what Tifft had originally claimed, but is of course, consistent with that claim.

In the meantime, there had been parallel developments on the quasar front too. One of us (GB) had noticed in 1968 that the redshifts of quasars and active galaxies are not smoothly distributed but show peaks at periods of 0.061. In 1969 he and Margaret Burbidge found a strong peak at 1.955 (1.96). This

Fig. 16.16. Bill Tifft who was the first to detect a periodic pattern in the redshifts of galaxies in a cluster. Credit: *Bill Tifft*.

finding was also hard to understand and most cosmologists felt that the result would disappear as more and more quasars were discovered. However, more peaks were found, and starting in 1971, Karlsson performed a mathematical analysis showing that there should be peaks in the redshift distributions at the following values of redshifts:

0.061, 0.30, 0.60, 0.96, 1.41, 1.96, 2.63, 3.44, 4.45,...

This series may look to be without any pattern. However, there is one. Add one to each number and then the series becomes geometric, with each term 1.277 times the previous one. The addition of 1 is natural since under a redshift z, the original wavelength is multiplied by $(1 + z)$. In the year 2000 one of us (GB) and Napier found observational evidence for the peaks at 2.63, 3.44 and 4.45.

The Karlsson result has also remained with increased data, and these red-shifts, uncannily, keep turning up in samples of quasars suspected of having anomalous redshifts. Even outside the conventional framework, there is no explanation for them. If one uses the variable mass hypothesis, then one has to assume that somehow there was a quantum effect controlling the mass of a particle which instead of growing continuously from zero at creation, jumps through a geometric series of values. Why? We do not know!

The tendency of cosmologists is simply to ignore those results which most believe to have arisen from improperly selected data or faulty analysis. Yet they have not been faced head-on and resolved one way or the other. This is somewhat ironic, as we have seen that most cosmologists would uncritically accept speculations about dark matter or dark energy for which there is no direct evidence; but which they need for the survival of their theory.

There is a maxim that scientists follow, and which has proved useful in the progress of science: *Trust no theory that has no observation to support it.* The attitude of the conventional cosmologist to these anomalous results on his/her doorstep seems to be the exact opposite: *Trust no observation that has no theory to explain it.*

As we struggle to make progress through the labyrinth of facts and speculations, we cannot help feeling that cosmology will not make real progress until it has resolved these claims for anomalous redshifts. If they all turn out to be wrong, we shall have renewed confidence in the framework based on Hubble's law and the expansion of the universe. If at least some of them turn out to be real, then we may be on the brink of a new breakthrough in cosmology.

17

Epilogue

As we did at the beginning, we reiterate that cosmology – the investigation of the real universe as we know it – is an observational science. This means that it has been driven by the observational discoveries that have been made, and to a large extent by the certainty with which major discoveries have been accepted, and the ways in which the leaders in the field at the time of discovery have treated the results.

Up to the early part of the twentieth century it was generally believed that the universe (which really was assumed to extend only as far as the Milky Way) was static.

A huge step forward came with the discovery that the universe is largely made up of individual galaxies sometimes arranged in clusters and groups all moving away from each other – the expanding universe – a discovery largely due to the astronomers led by Hubble at the Mount Wilson Observatory.

But already in making these discoveries extrapolation from observation to theory was being made. The interpretation of the apparent magnitude–redshift relation as being due to expansion, and the relation to the expanding solution of Einstein's equation and thus to a cosmological model which was generally believed by the world at large by about 1930, was not the only possibility. What was ignored was the idea that the redshifts could have a different explanation. Soon after the discovery of the redshift–apparent magnitude relation by Hubble and Humason several prominent physicists suggested that the redshifts had a different explanation – the so-called 'tired light' hypothesis in which it is supposed that the photons from a distant source lose energy as they travel to us. While this idea has been revived from time to time, there are severe physical objections to it, but it is clear that at the time that it was first suggested it was highly unpopular compared to the view that the results could be directly connected to Einstein's theory.

The other possibility was that although the universe is expanding this does not necessarily imply a beginning. This view did not surface until 1947 when Gold suggested it in the arguments he was having in Cambridge with Hoyle and Bondi. Gold suggested that while the universe is expanding it remains unchanged, i.e. it remains in a steady state. They were sceptical, but finally concluded that Gold's suggestion was viable. This led them in 1948 to propose the classical steady-state cosmology.

As we described in Chapter 10, this theory was immediately subjected to intense scrutiny, and most astronomers were critical of it, in many cases because they had already accepted, and believed in an evolving universe. There were genuine difficulties in explaining all of the observations in the framework of the steady state, but in addition the idea was generally disliked.

By the late 1960s discussion of the steady state had largely ceased while the conventional big-bang model went from strength to strength.

There was a second way of answering Gold's question – could one have a universe without a beginning which still acccommodates the Hubble relation? This possibility has been revived recently, in the form of a model of the universe which is oscillating. With Fred Hoyle, we developed this model in the 1990s and called it the *quasi-steady-state cosmology*. It is a cyclic universe, as described in Chapter 15.

Without making more assumptions it is impossible to explain the observed universe using any one of these three approaches, the evolutionary scheme – the big bang, the classical steady state, or the oscillating universe. In each case some theory which goes way beyond what we have so far established in physics, must be assumed.

Most effort has been expended on the big bang, since it has attracted the most attention. We begin our concluding remarks with that model.

Limitations of the big bang

All of the problems of this model are concentrated on the initial phase – the early universe. There must be a beginning but we have no understanding at all what this really means; remember that the creation process, whatever that is, also involves the creation of the laws of physics! Times as short as 10^{-43} sec after the beginning are used but we have no operational meaning to attach to such a short time scale. The number arises from the quantum theory of gravity. As yet we have no such theory on a firm footing and even when we do (if we ever do!) we will not be able to identify any phenomenon either in the laboratory, or in the cosmos, where this time scale appears. In short, the time scale is highly esoteric.

The speculations do not stop there. At around 10^{-37} second, the universe is supposed to have passed through the GUT era. At present particle physicists have not homed in on a specific grand unified theory, although most believe in the holy grail of such a theory. During this era the universe is believed to have undergone inflation, although big-bang cosmologists do not have a well-established, and universally accepted framework to describe it. Although claims are made that findings of WMAP confirm inflation, how do you 'confirm' an idea for which there is no well-accepted framework?

Sometime during this era also a 'decision' was made in favour of 'matter', vis-à-vis 'antimatter'. Particle physics developed in the 1960s and 1970s on the firm belief of a clear symmetry between matter and antimatter. For particles like electrons and protons, counterparts, positrons and antiprotons exist. The antiparticle has opposite electric charge to that of the particle; but more importantly, when brought together, a particle and its antiparticle annihilate each other completely, producing pure radiation in the process.

However, if the big bang started with perfect symmetry of this kind, there would be no particles left. So a scheme has to be devised wherein this symmetry was destroyed early enough, presumably in the GUT era. How this happened and the universe was left with a surplus of matter, today belongs in the realm of speculation. A well-established GUT may throw light on the problem.

However, this issue also has a bearing on *why* today the cosmic microwave background has a temperature of 2.7 K. At this temperature there are some 400 photons per cubic centimetre. How many baryons per cubic centimetre do we have? In the big-bang cosmology, as we saw earlier, this number is crucial. For, the parameter which has to be *chosen* in this theory, is the ratio of photons to baryons that will allow the synthesis of helium and deuterium (the latter more sensitively) in primordial nucleosynthesis in this early universe. This number was deliberately chosen by Gamow in his early investigation because he concluded (using observations that turned out to be incorrect) that the bulk of the helium in the universe must have been made there. Modern calculations tell us that if the number of baryons exceeds a certain limit, the observed quantity of deuterium could not have been made in that process. Which is why, the big-bang cosmology requires the bulk of dark matter to be non-baryonic.

What was not known then (in 1950) and was not established until much later, was that the energy released in the burning of hydrogen into helium (in stars) will give rise to the energy density of black-body radiation at $T \simeq 2.75$ K, almost exactly the observed temperature of the microwave background radiation (*see* Chapter 12). Since this comes from observed quantities it is an extremely important number.

The other problems of the big-bang theory are all associated with the making of galaxies. How to make self-gravitating lumps out of an initially smooth medium? The early ideas were to start with density fluctuations which will grow as the universe expands. These are believed to have had a quantum origin.

However, in the early days it soon became clear that in an expanding environment the fluctuations will not grow without seriously disturbing the smoothness of the cosmic radiation background. Thus, to solve this difficulty a new form of matter – non-baryonic matter – originally hot or warm and now cold, is assumed to be present. This has the convenient feature that it provides gravitational attraction but has none of the inconvenient properties of normal matter. In particular, it does not interact with radiation. A large amount of work is now being carried out making numerical simulations of galaxy formation dominated by cold dark (non-baryonic) matter and it is claimed that the large-scale structure of the observed galaxies in the universe can be explained in this way.

Most attention has been paid to studies of microwave background radiation, which following Gamow and his associates is thought to be the expanded remnant of the initial hot gas ball that came with the creation. These studies are considered important because the radiation background is believed to carry imprints of several primordial processes and events which astronomers cannot observe directly.

In this picture matter and radiation expanded together but at different rates, and at a redshift of about $z = 1000$ they had effectively decoupled. However the imprint of the initial density fluctuations out of which the galaxies formed was left in the form of very tiny distortions of the otherwise smooth radiation background. These distortions are being intensively studied and it is claimed that this model works well. It is claimed that, even in the latest observations there are indications that there was a very early inflationary period.

There are significant weaknesses in the whole of this observational programme. To start with, it has been planned and carried out by large groups of observational cosmologists who have started with the assumption that the standard model is correct. In any good scientific experiment there is nearly always someone who is highly sceptical of the theory and would very much like to show that it is wrong. This is not true here. Thus when anything is found that is not expected, the tendency has been to invoke a new parameter which can be incorporated into the theory just as the counter-Earth was invoked by the Pythagoreans. With this approach, no observations can ever disprove the basic framework of the theory. The possibility that a quite different explanation might fit the data equally well is of course, not considered. The process is strongly reminiscent of the epicyclic approach of the Greeks.

Thus the standard big-bang model cannot explain what we see unless we incorporate these epicycles:

- Choose a value for the initial baryon/photon ratio
- Assume that there are initial density fluctuations prior to the big bang
- Assume that an inflationary phase happened
- Assume that the bulk of the mass-energy in the universe is in the form of dark baryonic matter, dark non-baryonic matter and dark energy in stipulated ratios
- Assume that the phenomena described in Chapter 16 have no bearing on the present state of the universe. That is assume that redshifts can be interpreted in the normal way.

Thus, this is where we are at – a big-bang universe with inflation (for which there is no real theory), non-baryonic matter (for which there is no evidence) and dark energy (for which there is no understanding).

Alternative cosmologies

At this stage, a big-bang supporter may turn around and argue, that with all its speculations and other defects the big-bang cosmology is the only working theory of cosmology. Indeed, this argument is often made, implying that there are currently no viable alternative models in the field. Certainly, the classical steady-state theory of Bondi, Gold and Hoyle which played a rival's role in cosmology between 1948 and 1965, is no longer a viable alternative. Its failure to explain the origin of the cosmic microwave background, the origin of light nuclei and the observed evolution in the universe are reasons strong enough to rule it out as a possible contender.

We next refer to the so-called *relativistic MOND theory* which we have not discussed before. From the early days after the observational arguments for the presence of dark matter were made, Mordekai Milgrom suggested that perhaps there was no dark matter, but that we were wrong in supposing that Newton's laws of dynamics and gravity hold over very large scales. He argued that we see apparent stability in galaxies and groups and clusters not because there is dark matter present, but because their gravitational dynamics does not follow the pattern we expect when we accept Newton's laws. He used a modified Newton's law. This theory called MOND was published in 1983 (Modification Of Newtonian Dynamics).

While the use of MOND means that many puzzles normally requiring dark matter can be understood without this, it was not considered a real theory until 2004, when J. Beckenstein, R. Sanders and M. Milgrom devised a relativistic version of MOND. It is now claimed that this theory can provide a good fit to the data on microwave background fluctuations found by WMAP.

We feel that this approach deserves further scrutiny; particularly one should examine, how it deals with other cosmological data not related to dark matter.

We next come to the quasi-steady-state cosmology (QSSC) which we described in Chapter 15.

The main difference between this theory and the standard model or the MOND version is that there was no beginning, i.e., all of the mass-energy in the universe is due to explosive creation in regions of high density (the centres of galaxies) and this creation process is also responsible for the 'bounce' as the universe is contracting, so that it never evolves into a very dense phase. The universe also slows down as it reaches a maximum in the cycle leading to the contracting phase.

The underlying theory is based on the Hoyle–Narlikar theory of 1964 in which it is shown at the classical level that at very small distances, classical Newtonian attraction is overwhelmed by a repulsive field – the *C*- (creation) field. It is through this field involving negative energy, that new matter is created.

In view of the large amount of observational evidence that the centres of many galaxies are sites of great activity with the ejection of matter and energy in explosive forms, and the ejection of coherent objects, quasars and also compact galaxies, it is natural to suppose that it is in these highly active nuclei that matter and energy are being created. Thus we are incorporating here the cosmological ideas of the 1950s and 1960s due to Viktor Ambartsumian.

As explained in Chapter 15, the QSSC has been applied to several cosmological phenomena and has prima facie demonstrated itself to be a serious contender in the field. We look at this theory alongside the big-bang theory by way of comparison.

The first is the classical view that there was a beginning and that all of the important properties of the universe were largely determined in the first few moments after the initial outburst. This is the most popular view.

But it has major problems. None of the quantities that are required to explain this very complex universe have a basis in any substantial theory. Inflation is invoked, the initial ratio of baryons to photons is chosen, and large amounts of non-baryonic dark matter for which there is no evidence at all, and dark energy, are also required. Moreover, all these inputs are speculative with no base in proven physics.

Part of this is improved if we accept the relativistic MOND theory, but at a price.

Also all of the effects seen in active galaxies are explained in the standard model by attributing them to matter ejected from massive black holes in the cluster of galaxies. When looked at in detail this is a classical bootstrap argument. In our view it is energetically impossible: that is, the model imposes unrealistic demands on the basic black hole scenario.

And of course the ejection of quasars from galaxies – the formation of new galaxies – is rejected out-of-hand by those who believe in that standard model. They clearly believe (and indeed, have to believe) that the data are being misinterpreted.

While there are large effects which they are happy to ignore, the very tiny effects associated with perturbation of the cosmic microwave radiation background are absolutely believed, not surprisingly, by the people who set out to find them.

The alternative view is that there was no beginning and we live in an oscillating universe. The major weakness of this theory is that we have not been able to understand in detail the physics of the creation process. The circumstantial evidence for the scheme is good – for example, the prediction that the temperature of the microwave background was determined by the synthesis of helium from hydrogen is remarkably close to reality.

Undoubtedly the greatest difficulty with this model is due to the fact that very few professional cosmologists know about it, and very few work on it in an era when all of the publicity is on the other side. However, whatever picture finally emerges as a viable model of the universe will have to face the issues raised in Chapter 16, for in our view, today the most basic question to be answered is:

Do we really understand the nature of the redshift?

Index

Abell, George 146
absolute scale of temperature 174
absorption lines 80
abundances 154, 155
  of light nuclei 231, 232, 238–241, 247
acceleration 25, 26
Adams, John Couch 32–34, 47
age of the universe 123, 148
Airy, George 34
Alcatel-Lucent 177
Alpher, Ralph 176, 178, 185, 186
alternative cosmologies 277–279
Ambartsumian, Viktor 216, 235–238,
  252, 278
Ampère, André Marie 189
Andersen, Hans Christian 207
Andromeda Nebula 57, 58, 60, 70, 252
anomalous redshifts 250
  and alignments 259–262
  in close pairs and groups of galaxies
  251, 255
  from VMH 266
  in quasar-galaxy associations: see alignments
antimatter 132, 226, 275
anti-particles 153
apparent brightness 42
apparent magnitude 84
Archives Lemaître, Université Catholique
  102
Aristarchus of Samos 14, 15, 39, 41
Aristotle 11, 12, 21, 104
  hypothesis of natural motion 11
  hypothesis of violent motion 12
arithmetical series 268
Arp, H. C. (Chip) 253, 255–262, 268, 269
Arp 220 252, 257
Arp–Hazard triplets 261
Aryabhata 15–16
  Aryabhatiya 15
astrologers 12
astroparticle physics 187–194, 205
Australia Telescope National Facility 114
axiom 94

B^2FH 170, 185
Baade, Walter 114, 116, 141, 165
  bet with Minkowskii 116–118
background X-rays 175
barred spiral galaxy 70
baryons 132, 153
baryon number conservation 132
baryon number violation 133
baryonic (normal) matter 183, 222
Beckenstein, J. 277
Bentley, Richard 90, 92
beryllium 155
Bessel, Friedrich 39
  measurement of parallax of *61 Cygni* by 39
beta decay 189
Bethe, Hans 157
Bhagavatam 1
big bang 101, 104, 121–124, 239, 274
big crunch 101, 104
black-body radiation 173–174, 179, 184, 230, 241,
  242
black hole 209–210, 233
  strong attraction near 234
black money 209
blueshift 81, 125, 248, 264
Bohr, Niels 269
Bolton, John 114
Bondi, Hermann 87, 119, 120, 123, 128, 134, 137,
  184, 230, 242, 263, 274
boron 155
Bowen, Taffy 144
Bradley, James 39
  Aberration measurement of *Gamma Draconis* by
  39
Brahma 1, 2
Brahmanda 2, 3
Brown dwarfs 214
Burbidge, Geoffrey 118, 165, 170, 184, 185, 239,
  242, 270, 271
Burbidge, Margaret 165, 169, 170, 185, 252, 270

C- field 131, 232–235, 247
3C 264 236

3C 273 150, 151, 236
  occultation by Moon of 150–151
CCD 44
Cameron, Al 169
carbon 155, 162–164, 168
CDM (cold dark matter) 218, 222, 227
  Candidate particles of 218
Central fire 9
Cepheid Variables 66–68
  period–luminosity relation for 67, 68
CERN 117, 188, 191
Challis, James 34
Chandrasekhar, Subrahmanyan 110, 166
Chandra X-ray Observatory 115, 236
chemical elements 153
  atomic charge of 154
  atomic weight of 154
  building of 154–159
Chinese mythology 7
Clerke, Agnes 59, 60
closure density 106
clusters of galaxies 171, 215, 237, 238
closed space 197
CMBR screen 193
CN cycle 158, 159
COBE satellite 178
  spectrum of radiation measured by 179, 231
  inhomogeneities of radiation found by 181, 182, 231
colliding galaxies 116, 117
Coma cluster 215, 269, 270
concordance in cosmology 228
Copernicus, Nicolaus 15, 16, 17, 20, 47
Cosmic hierarchy 3
cosmic radiation 173
cosmic rays 153
  energy of 172, 175
cosmological constant Λ or λ 97, 100, 148, 184, 222, 225, 235, 247
  volte-face on 222
  with negative sign 235, 246
cosmological force 97, 218
cosmological models 89, 97, 98
  de Sitter's universe 98
  Einstein's universe 92–98
  Friedmann (big bang) models 98–101, 148
  Hoyle–Narlikar models using C-field 131, 133, 134
  Newton's universe 90–92
  open and closed 104–109
cosmological principle (CP) 103, 104, 124
cosmology 89
Cotsakis, Spiros 15
Cotton, W.D. 236
Counter-Earth 9, 10, 205, 217, 276
counting of galaxies 108–111
counting radio sources 140–147
Crab Nebula 57, 166, 167, 169, 244
  explosion recorded by Chinese observers 57
  explosion recorded by American Indians 57
Crab pulsar 244

Crab supernova 244
creation of matter 123, 128–131, 232, 279
  continuous 125–128
critical density 104, 105, 198, 206
Curtis, Heber 60, 62
curvature of space *see also* open, closed and flat space
  negative 106–108
  positive 106–108
  zero (flat space) 106–108
cyclic universe 274
61 Cygni 41
Cygnus A 114–116

Dadhich, Naresh 15
dark ages 228
dark energy 183, 218–223, 223–225, 227, 231, 247
dark matter 183, 214, 231
  composition of 214–215, 216–218
  in clusters of galaxies 215–216
  in spiral galaxies 210–212
  in QSSC 249
dark baryonic matter 183, 217
darkness of night sky 75–78
Das, Prashanta K. 264, 268
De Revolutionibus 17
de Sitter, William 98, 223
de Vaucouleurs, Gerard 146
debate between Shapley and Curtis 60–63
deceleration parameter 148
decoupling of radiation from matter 226
Delta Cephei 245
density fluctuations 226
density parameter Ω 105, 222
Deshpande, Kshitija 110
deuterium 239
  limit on primordial 217
deuteron 160
2dFGRS Team 146
Dialogue on the Two World Systems 20
Dicke, R.H. 176, 177
Dingle, Herbert 134
Doppler 80, 81
Doppler effect 80–81, 82, 264

Eagle Nebula 43
Eddington, Arthur 110, 120, 122, 156
Einstein, Albert 8, 92, 93, 96, 100, 148, 190, 218, 223, 265
  gravity theory of 47, 230, 257, 273
  relation $E = Mc^2$ 157
Einstein's universe 92–98
electromagnetic field 189
electromagnetic wave 113–114
electrons 153, 154, 185
electro-weak theory 190
elliptical galaxy 71, 72
emission lines 80
Emperor's New Clothes 207–208
end of physics 8, 190, 191
epicycles 9–27, 225

epicyclic theory 12–14
epicycles of modern cosmology 224, 276
epoch of last scattering 180–183, 193, 194
equilibrium process 165
erg 172
Eta Aquile 248
Euclid 94
ESO (European Southern Observatory) 70
evidence for dark matter 208–216
exact solution 205
expanding universe 78, 84, 89, 98–103, 104, 272, 273
explosive events 235–238

face of God 182
Faraday, Michael 129, 189
Feretti, L. 236
field 129–130
  of negative energy 129, 130–131
fine tuning 197, 198, 204, 223
fireball 230
flat rotation curves 210–212, 213, 223
flat space 197
flatness problem 195, 197, 204
flux density 141
formation of light atomic nuclei 226
Fowler, Willy 164, 165, 169, 170, 185, 263
Fraunhofer, Joseph von 80
  Fraunhofer lines 80
Frederick the Great 32
Friedmann, Alexander 98–103, 120
  Friedmann models 98, 101, 104, 105, 118, 149, 203

Galactic centre 51–55, 244
galactic disc 58–59
Galaxy 46, 50, 54, 57, 70, 114, 175
galaxy types 255 (*see also separately under different types*)
galaxy groups 927
Galileo Galilei 17–20, 24, 46, 47
  Inquisition on 20
  Proof of motion of the Earth by 37–39
  Use of telescope by 17–18, 19
Galle, Johann G. 33, 34, 209
gamma ray bursts 237
  burst source GRB 031203 237
Gamow, George 159, 160, 176, 177, 178, 185, 217, 230, 275
  predictions of relic radiation temperature by 178–179
general relativity 96, 97, 107, 205
  exact solution in 205
geocentric theory 12, 20
geometric series 269
geometry 93–97
  Euclidean 94, 95, 106
  non-Euclidean 94, 97, 106
George III, King of England 36
GeV (Giga electron volt) 191
Giovanni, G. 236

globular clusters 52, 53
Gold, Thomas 114, 119, 120, 123, 134, 137, 142, 145, 184, 230, 242, 274
gold sample of Type Ia supernovae 246
Goodricke, John 66
gravitational constant G 233
gravitational force 189
gravitational lensing 258
gravitational redshift 262–264
Gunn and Oke 149
GUT (grand unified theory) 190, 191–192, 199, 247
  GUT energy 192, 200
  GUT epoch 193, 196–198, 201, 204, 226, 275
Guth Alan 134, 200, 205
Guthrie, B.N.G. 270
Gutierrez, C.M. 254, 255

Hale, George Ellery 110
Hale Telescope 110, 111, 147, 150, 151
Halley, Edmund 30–32, 269
Halley's comet 30–32, 47
  in Bayeux Tapestry 31
  in the most recent visit 31
Hanbury Brown, Robert 150
Hawking, Stephen 190
Hazard, Cyril 150, 151, 261
HDM (hot dark matter) 218
  massive neutrino as 218
heliocentric theory 17, 20
helium 155, 159, 162, 164, 168, 239, 240–241
  made in stars 239, 240
Hercules cluster of galaxies 238
Herman, Robert 176, 178, 185, 186
Herschel, William 32, 35–37, 46, 50, 58
  Discovery of Uranus by 32
  Milky Way map by 36–37, 47, 50, 51
  telescopes built by 35, 36, 47
Hertz, Heinrich Rudolf 113
Hertzsprung–Russell diagram 163
Hipparchus 17
homogeneity 90, 103
horizon from Earth's curvature 195
horizon problem 194, 195–197, 204
Horse-Head Nebula 49, 116
Hoyle, Fred 101, 119–121, 128, 129, 131, 137, 142–145, 150, 165, 169, 170, 184, 185, 203, 230, 232, 239, 242–244, 263, 264, 274
  breakthrough in nucleosynthesis by 161–164
Hoyle: private collection 120, 170
Hoyle and Narlikar 150, 203, 264
  gravity theory of 265, 278
Hubble, Edwin P. 63, 64, 69, 72, 73, 78, 82, 83, 98, 100, 102, 103, 104–106, 107, 108, 110, 111, 120, 147, 148, 159, 230, 272
  measurements of distances of galaxies by 62, 63–68
Hubble's constant 147–148, 159
Hubble's law 82–86, 87, 115, 138, 142, 147, 250, 253, 254, 256, 258, 260, 272

Hubble bubble 131–134, 203, 204
Hubble Space Telescope Image Gallery 73
Humason, Milton 82, 100, 147, 159, 230, 250, 273
Hung, Ko 7
hydrogen 153, 155, 168, 180

Indian Institute of Astrophysics 31, 80
inertia 265
inflationary universe 134, 182, 200–203, 222, 223,
    224, 247
  era of 226, 276
  is it necessary? 205
  new models of 204
  lack of a definite theory of 205
inhomogeneity of large-scale structure 145
intergalactic dust 225, 231, 246–247
interstellar dust 48–49, 51
interstellar extinction 50, 52, 54
inverse square law of illumination 43, 44, 47
  pitfalls of 48–50
invisible clothes 222
iron group of elements 161, 165
irregular galaxy 73
isotopes 155, 239
isotropic source 44
isotropy 90, 103

Jain, P. 245
Jansky, Karl 111, 112–114, 141
Jansky (unit of flux density) 141
Jodrell Bank 114
Joshee, Deepak 13

Kant, Immanuel 56
  island universe hypothesis of 56, 60
Kaptyn, J.C. 51, 52
  Kaptyn universe 51–53, 60
Karlsson, K.G. 271
Kazanas, Demosthenes 200
Kelvin, Lord 178
Kelvin temperature scale 159
Kepler, Johannes 12, 20–22, 24
  three laws of planetary motion by 20, 23, 27, 210
Kitt Peak National Observatory (KPNO) 71
Kukudmi 1

Lambert, J.H. 56, 58
de Laplace, Pierre Simone 34
Lara, L. 236
large-scale structure 227, 276, 146
leaning tower of Pisa, 17
Leavitt, Henrietta 65, 66–68
Lemaître, Abbé 101–103, 120, 136, 230
Leverrier, Urbaine Jean 32, 34, 47
light year 137
Local Supercluster 146
Lopez Corredoira, Martin 254, 255
Lord's Bridge 114
Lovell, Bernard 114, 136
luminosity 42, 162
  of the Sun 45

M 32, elliptical galaxy 71
M 82, irregular galaxy 73
M 101, spiral galaxy 69, 211
Mach, Ernst 264–266
Mach's principle 265
Mackay, M.B. 151
Magellan 66
Magellanic Clouds 66
magnetic field 129
  in the Galaxy 175
main sequence 162, 163
Markarian 205, 255, 256
2MASS/UMass/IPAC-Caltech/NASA/NSF
  171, 215, 270
massive objects 233
mass–luminosity relation 229
Mather, J.C. 178
matter domination over antimatter 226, 275
Matthews, Tom 151
Maxwell, James Clerk 190
Mayall, Nicholas 250
McKeller, A. 175, 177, 178
Mecanique Celeste 119
merging galaxies 252
Merrill, Paul 169
Messier, Charles 56
  Messier catalogue 56–57, 69
metallic whiskers 243–246
microwaves 172
microwave background 171, 173, 175, 231, 232,
  247, 275, 276
  first detected 175–176
  predicted as relic radiation 176, 177, 185
  conjectured as relic starlight 184
  fluctuations of 225, 245–246, 247, 276; *see also*
    COBE, WMAP
Milgrom, Mordekai 214, 277
Mills, Bernard 114, 142
Mills cross antenna 114
Milky Way 36, 46, 51, 54, 146, 252
minibang 234
mini-creation event (MCE) 234–235, 239
Minkowski, Rudolf 114, 116
MOND (modified Newtonian dynamics) 214,
  277–278
monopole problem 195, 199–200, 204
Monte Carlo programme 145
Motion of the earth 37–39
  Aberration resulting from 39
  Galileo's proof of 37–39
Mount Wilson Observatory 51, 61, 82, 107, 147,
  165, 273
  60-inch reflecting telescope in 61, 64
Mount Wilson and Las Campanas Observatory
  64
Mullard company 143, 144

Napier, Bill 270, 271
Napoleon 34
Narlikar, Jayant 131, 143–145, 146, 203, 264, 266,
  268

NASA 43, 49, 51, 57, 64, 219
  Hubble Heritage 211, 251
NASA/CXC/M. Weiss 237
Nature of the universe 121
near-black hole (NBH) 233, 234, 235
  creation near 234
nebula 42
Neptune, discovery of 32–34, 208, 209
neutrinos 153, 164
neutrons 153, 164, 168, 239
New General Catalogue 69
Newton, Isaac 12, 24, 25–29, 47, 78–90, 91–93,
    223
  apple tree of 28–30
  law of gravitation by 25, 27, 32, 34, 47, 88, 213
  laws of motion by 26, 34, 265
  magical years in the life of 28
  tests of the laws by 29–30
  universality of the laws by 28–29
Newton's universe 90–92
NGC 1365 70
NGC 3079 257
NGC 3628 257
NGC 3810 256
NGC 4038–39 252
NGC 4256 257
NGC 4258 260
NGC 4319 and M 205 255
NGC 4676 252
NGC 5128 252
NGC 5548 257
NGC 6212 and 3C345 257
NGC 7603 254–256, 268
NOAO/AURA/NSF 238
non-baryonic dark matter (NBDM) 183, 214,
    217–218, 226, 231, 246, 247, 275, 276
Norse world tree 5, 6
Noto Keith 50
nucleus of an atom 153
nuclei of galaxies 134

Oke, Beverley 140, 151
Olbers, Heinrich 75, 76, 78, 241
Olbers paradox 75, 86, 87, 125, 240
onion-skin structure 164, 166
Oort, Jan 54, 114
open space 197

Palomar Observatory 148, 151
Palomar Observatory and California Institute of
    Technology 83
parallax 39, 41, 47
  use in distance measurement 40–42, 63
Paranjpye, Arvind 29, 36, 52, 171
Parkes 151
parsec 147
particle horizon 109, 194, 195
past light cone 137
Payne, Cecilia 156–157
peculiar galaxy 73
Penzias, Arno 177, 178, 184–186, 230

perfect cosmological principle (PCP) 121, 124–
    128, 138
periodic redshifts 268–272
phase transition 200
Philolaos 9
photons 87, 153
physics of stars 228–229
Pigott, Edward 66
Planck constant h 233
Planck particle (PP) 233
planet 12
planetary nebula 168, 169
Pope on cosmology 136
Popper, Karl 128
positrons 153
postulate 94
primeval atom 103, 120, 230
primordial nucleosynthesis 159–161, 179, 186,
    188, 217, 223, 231
Princeton University Physics Department 176
Proctor, R.A. 58
proper motions 61–63
protons 153, 154, 164, 239
proton–proton chain 157, 159
Pryce, Maurice 131
Ptolemy 14, 17
Pujari, Anagha 65, 156
Pythagoras 9
Pythagoreans 9, 10

quantum mechanics 229
quantum theory of gravity 192, 233
  era of 226, 247, 274
quasi-steady-state cosmology (QSSC) 230, 232–
    235, 240, 252, 274, 278–279
  blueshifts in 248
  burnt out stars as dark matter in 241
  comparison with the standard model
      247–249
  cycles of 240
  long-term time scale of expansion 232
  relic radiation from burnt out stars in 241–246
  scale factor of 232
  short-term time scale of oscillation 232
  synthesis of light nuclei in 239
  very old stars in 248–249
quasi-stellar objects (quasars, QSOs) 150–152,
    236, 237
  aligned triplets of 261
  ejection from galaxies 279

Ra, Sun God 4–5
radio astronomy 112–114
  and cosmology 118
radio jets 236
radio sources 114–118, 237
  optical identification of 115–116, 141
rapid process 169
Reber, Grote 112–114
red giant 165, 168, 169
redshift 81, 125, 264, 271

redshift–magnitude relation 149, 221, 230, 245, 247, 273
repeatability of experiments 185
    lack of in standard model 228
relic starlight 184
resolution 51
Revati 1
Reynolds 72
right hypothesis 10
Ring Nebula 57, 58, 169
Robertson, H.P. 103
rotation speeds of neutral hydrogen clouds 212–213
rotations of spiral galaxies 210–212
Royal Astronomical Society 144, 145
Russell, Henry Norris 156
Rutherford, Ernest 122
Ryle, Martin 114, 142–145, 150
Ryle–Hoyle controversy 143–147

Salam, Abdus 190
Salam–Weinberg approach: *see* electro-weak theory
Sandage, Allan 140, 148, 149, 151, 219, 250
Sanders, R. 277
Sato, Katsuhiko 200
scattering of light 172, 179, 180
Schmidt, Maarten 151
scientific hypothesis 10
selection effect 71
Shah, R. 245
Shapley, Harlow 52–54, 60, 62–68
Shimmins, A.J. 151
Shiva 1
singular epoch 187
Sky and Telescope 121
Slipher, Vesto Melvin 81
slow process 169
Smiley, Patricia, NRAO 112
Smoot, George 182
Souradeep, T. 245
spectroscope 79
spectrum 79
speed of light c 233
spiral arms 54
spiral galaxy 54, 69–71, 211
spiral nebulae 60
standard candle assumption 45, 48, 221, 231
standard model (*see also* big bang) 226–228
    limitations of 274–277
    schematic view of 227
static universe 90
statistical mechanics 187, 193
steady state 121, 124, 136
steady-state theory 118, 119–131, 137, 138, 145, 148, 149, 204, 232, 274, 277
    accelerating expansion predicted by 148, 149, 183, 219, 222
    age distribution of galaxies in 139
    horizon in 126
    response to by academic community 134–135
Stebbins, Joel 137, 139

Stebbins–Whitford effect 137–140
stellar evolution 167–170, 247
stellar nucleosynthesis 186, 247
Stephan, E.M. 252
Stephan's quintet 252–254
strong force 189
Sulentic, Jack 256, 269
Sun 163
superclusters 145, 146
supernovae 165, 168, 220, 243
    light curve of 167, 220
    redshift–magnitude relation for 221, 231, 247
    Type Ia 219, 220–221, 224, 246–247
supernova cosmology watch 221
supercooled steam 201
Sydney 151
symmetry 11, 12, 104, 124
synchrotron radiation 117
Syntaxis 14
synthesis of heavier nuclei 164–166

Telescope in Education 69
tidal interaction 251
tidal phenomenon 38, 47
Tifft, Bill 269, 271
time–temperature relationship 186
tired light hypothesis 273
Tolman, R.C. 108, 120
tuning fork diagram of galaxy types 72
twenty-one centimetre wavelength 211, 212
Tycho Brahe 20, 21

unified field theory 190
universe: accelerating or decelerating 148–150
    accelerating 148, 183, 219, 221
    decelerating 148, 219, 221
untested in the laboratory 228
Uranus planet 32, 209
    Discrepancy in the orbit of 32–34

van Maanen, Adrian 60–61, 63, 73
van Maanen's observations 60–63, 73
variable mass hypothesis (VMH) 264–268
velocity–distance relation 84, 100, 102
Venturi, T. 236
very early universe 187
Virgo Cluster 146
virial theorem 216
Vishnu 1
voids 146
Vorontsov–Velyaminov 254
VV 75 172

Walker, A.G. 103
wavelengths of coloured light 79
weak force 189
Weyl Hermann 103
Weyl's postulate 103
Whaling, Ward 164
White hole 234

Whitford, Albert 137, 139
Wickramasinghe, Chandra 243, 244
Wien, Wilhelm 174
    Wien's law 174
Wilkinson, David 176, 177, 182
Wilson, Robert 177, 178, 184–186, 230
WIMP 217
WMAP implications 176, 182–184, 231, 245, 275

wrong hypothesis 9

Yang 7
Yerkes Observatory 61
Yin 7
yugas 1

Zwicky, Fritz 84, 165